D1569670
LAKE COUNTY
Birch Lake
Stoney River
Greenwood Lake
TACONITE HARBOR
Erie Mining Company Railroad
Railroad
Range
Company
DAN
Duluth & Iron
Railroad
Highway 61
SILVER BAY
BEAVER BAY
LITTLE MARAIS
GRAND MARAIS
Lake
Superior
TWO HARBORS
APOSTLE ISLANDS
LA POINTE
ONTONAGON
IRON RIVER
ASHLAND
IRONWOOD
BESSEMER
HURLEY
WAKEFIELD
GOGEBIC RANGE
MELLEN
ONSIN
MICH.

Publications of the

MINNESOTA HISTORICAL SOCIETY

PIONEERING WITH TACONITE

By E. W. Davis

MINNESOTA HISTORICAL SOCIETY *St. Paul 1964*

Library of Congress Catalog Card Number: 64-64494

Photographs in this book which are not specifically credited were generously supplied by the Mines Experiment Station of the University of Minnesota and the Reserve Mining Company.

Foreword

THE AUTHOR of this book is a modest man and a stubborn one. Because of the first quality, the reader may emerge at the end of this volume with no really clear idea of E. W. Davis' pervasive contributions to the development of taconite processing. Without the second quality, there would have been no book to write, for Mr. Davis stubbornly persisted for over forty years in his efforts to perfect a successful method of processing the magnetic taconite on Minnesota's Mesabi Range. Minnesotans refer to him as the "Father of Taconite" or as "Mr. Taconite," and both appellations are accurate.

Edward Wilson Davis was born in 1888 at Cambridge City, Indiana. He completed his college work at Purdue University, obtaining a bachelor of science degree in 1911 and a degree in electrical engineering in 1918. Meanwhile in 1912 he had joined the staff of the University of Minnesota as a mathematics instructor in the School of Mines. His association with the university was to continue for forty-three years, as he became professor and superintendent of the Mines Experiment Station (1918–39), and director of the station from 1939 until his retirement in 1955. Some of the events in that long career are chronicled in these pages — the beginning of his work on taconite in 1913, his association with the Mesabi Syndicate and the Mesabi Iron Company which attempted taconite processing from 1915 to 1924, and his work as a consultant to the Reserve Mining Company from 1955 to the present.

Over the years Mr. Davis has authored many technical bulletins, papers, and articles dealing with phases of iron ore beneficiation, and he is the holder of at least eighteen patents. Among the numerous awards showered on him are honorary degrees of doctor of engineering from Purdue, of doctor of science from the University of Minnesota, and the Richards Award of the American Institute of Mining and Metallurgical Engineers for 1955. While the language used in conferring such honors is frequently exaggerated and flowery, that of the regents of the University of Minnesota in 1956 states only the simple truth. They spoke of him as "a symbol of what can be achieved by a scholar in action." They also referred to Mr. Davis as an "unyielding advocate, year in and year out, of the need for developing methods to concentrate Minnesota's abundant supply of intermediate grade ores." Characteristically, Mr. Davis himself says, "I was a pest on the subject."

Of all the honors received by this white-haired but youthful modern pioneer, perhaps the recognition which means the most to him came from the industry he fostered. In 1953 the Reserve Mining Company named the world's first large taconite processing and pelletizing plant at Silver Bay, Minnesota, the "E. W. Davis Works." At that time, the company presented Mr. Davis with a sumptuous, leather-bound, hand-lettered volume richly illuminated in gold leaf and red. The volume is one of his most treasured possessions.

Readers may also be interested in knowing that Mr. Davis' agile mind recently pioneered yet another field of research. Watching from the window of his home in Silver Bay in the summer of 1960, Mr. Davis saw several scuba divers disappearing and reappearing from the waters of Lake Superior. Later he learned they were seeking the hulk of a ship believed to have sunk there many years before. Coupling this new information with his long interest in the fur trade which flourished in northern Minnesota in the eighteenth and nineteenth centuries, Mr. Davis wondered if divers might not recover trade artifacts in the waters along the Minnesota-Ontario border. As a result, he set off with three scuba divers. They returned with a unique nest of seventeen graduated brass trade kettles found at the foot of a rapids. They presented the kettles to the Minnesota Historical

Society, of whose executive council Mr. Davis is a member. The society then launched a large-scale search for additional items — a search which has now widened to include the Royal Ontario Museum and many other organizations. Mr. Davis' idea that artifacts might be preserved in the cold fresh waters of the area proved sound, and in the years since 1960 thousands of specimens have been brought to the surface to enrich museum collections and enhance the historian's knowledge of Minnesota's first great industry — the fur trade.

This book grew out of a conversation in 1960 in which I, as the Minnesota Historical Society's associate editor, asked Mr. Davis to set down for the society the story of his work with taconite. After four years of intensive labor, he has produced more than that — a careful, annotated account which attempts for the first time to record the development of taconite processing, a development which has revolutionized iron and steel production in a few short years.

June Drenning Holmquist

MINNESOTA HISTORICAL SOCIETY
ST. PAUL, MINNESOTA
July 13, 1964

Preface

THIS BOOK was written to record the history of the development of taconite processing as I saw it and lived with it through its ups and downs for fifty years. During that time, it slowly and falteringly emerged from the Mines Experiment Station's laboratory at the University of Minnesota to become a thriving industry.

Along the way I was privileged to know and work with many competent, energetic, and generous people. Without them, taconite would still be only another rock in the earth's crust. In the pages that follow, the reader will meet some of these people, but there were many more who made valuable contributions, some with but a single word, others with long years of constructive effort. I hope that to some extent this book will repay all of them for their efforts and serve as a reminder that they are not forgotten.

This record could not have been assembled without the friendly assistance of many persons — more than I can possibly name here. Miss Mildred R. Alm of the Mines Experiment Station provided invaluable help, and always — well, almost always — was able to find for me obscure bits of information that I thought I remembered in a hazy sort of way. Henry H. Wade, my long-time associate at the university, and Harold H. Christoph, checked and rechecked facts and figures and made valuable suggestions, as did William K. Montague, a university regent and himself a versatile author. George M. Schwartz, for many years director of the Minnesota Geological Survey, was most helpful, as were my dear friends, Dale and Gerry Shephard, experts on the simple way of doing difficult tasks.

I am grateful also to many members of the Reserve Mining Company's staff: Edward Schmid, Jr., director of public relations, and his assistants; George Reid for preliminary sketches and maps; Mike Tourje for special photographic work; Tess McCarty, Art Samuel, Bob Lee, Larry Edwards, and a host of others who were so cheerfully helpful. My thanks also to the people in the public relations departments of the Armco and Republic Steel corporations and others who supplied photographs and read the manuscript for errors and omissions. Unless they are otherwise credited, all the photographs in this book were supplied by the Mines Experiment Station and the Reserve Mining Company. The overall design of the book is the work of Alan Ominsky of Minneapolis.

The staff of the Minnesota Historical Society deserves my special thanks, particularly those members of its publications department who must have struggled mightily with my writing and spelling. June D. Holmquist, my meticulous editor and sometimes slave driver, can now, I am sure, qualify as an expert on taconite processing. She is one of the few women to have had a personally conducted tour through the E. W. Davis Works at Silver Bay.

And last, but most important, I want to acknowledge the help of my good wife, who spent many hours searching old files, kindly made available to us by the St. Louis County Historical Society at Duluth, for elusive bits of information, and who was most helpful in finding papers that were misplaced after the friendly Finnish lady who came in to clean insisted on dusting my work table.

To all of these, and the host of others who helped in any way, I gratefully express my sincere appreciation.

E. W. Davis

SILVER BAY, MINNESOTA
June 30, 1964

Contents

List of Figures

This book is dedicated to the memory of three men whose vision and foresight gave taconite the breath of life: CRISPIN OGLEBAY, CHARLES R. HOOK, *and* ELTON HOYT II

And to two men who had the courage and the ability to make it a mighty industrial giant: CHARLES M. WHITE *and* WEBER W. SEBALD

Introduction

TACONITE is the name given by geologists to a type of hard rock containing fine particles of iron ore. The word was first used by Newton H. Winchell, a Minnesota geologist, who in 1892 applied it to the magnetic iron formation on the state's eastern Mesabi Range. Taconite is not useful as it comes from the mine, but the fine particles of iron embedded in it can be removed and made into excellent ore, which when smelted produces high-quality iron and steel more economically than any known natural ore. The particles may be magnetite (Fe_3O_4) or hematite (Fe_2O_3), magnetic or nonmagnetic, and as fine as dust or as coarse as beach sand. The one characteristic common to all taconite, however, is its low iron content, making elaborate and expensive processing necessary before it can be smelted.[1]

Iron has been known and used by man since the beginning of recorded history, and it is still his most useful and versatile metal. But the metal which is produced by smelting — heating iron ore under controlled conditions — bears no resemblance to the ore from which it is made. Natural ore is usually soft and brittle and shades in color from gray or brown through red to almost black. As it comes from the ground, it is an oxide (a chemical compound of iron and oxygen) mixed with such other materials as silica, alumina, and magnesia. High-grade ore contains no more than 5 to 10 per cent of such earthy impurities; lean or low-grade ore contains from 10 to 20 per cent. In taconite, however, these impurities amount to from 50 to 70 per cent.

If the earthy impurities in iron ore exceed 15 or 20 per cent, it is usually more economical to remove some of this worthless material by mechanical means (called beneficiation) before the ore is smelted. In processing taconite at the present plants in northern Minnesota, the amount removed constitutes nearly two-thirds of the rock mined. The remaining one-third is then agglomerated and shipped as high-grade ore to Eastern blast furnaces for smelting.

The making of metallic iron from iron ore is known as smelting. In the modern blast furnace, earthy impurities are removed in the form of molten slag, and the oxygen, in combination with carbon, leaves the furnace as a gas. Such furnaces have evolved over the centuries from small beginnings. The first smelting furnace was little more than a charcoal fire pot ringed with stones. A few small pieces of iron ore were placed in the pot and covered with burning charcoal. A blowpipe held in the mouth provided the necessary blast of air, and the result was metallic iron in the form of a spongy mass that could be pounded into a solid bar. This slow, laborious smelting was used by ancient man to produce small amounts of iron at least as early as 1600 B.C. and perhaps as far back as 4000 B.C.[2]

Air circulating through the furnace burden is the very lifeblood of smelting operations, and progress in the production of iron could be made only as rapidly as air-blowing equipment was evolved. A method, which involved inflated animal skins or bladders, was used to blow air into small smelting furnaces long before the Christian era. By treading first on one skin and then another, men forced air into the furnace. Using such equipment, the Greeks and Romans produced surprisingly large quantities of iron.

Over the years various other ingenious methods of producing greater quantities of air were devised, but it was not until the fifteenth century that the water wheel was connected to large mechanical bellows. This development made possible the operation of smelting furnaces three or four feet in diameter and fifteen or twenty feet high, capable of producing three or four tons of iron a day. The first smelting furnaces in the United States were of this type. The best known began operating on the Saugus River at Lynn, Massachusetts, in 1648.[3] About 1865, the steam engine came into general use, and steam-operated air pumps, or "blowing tubs" as they were called, were devised. By this time, too, coke was replacing charcoal as a smelting fuel, and much of the iron produced was molten, although some sponge was still being made. Limestone was added to make a slag, which absorbed the impurities in the ore and coke and flowed out of the furnace as a liquid.

By 1925 furnaces twenty feet in diameter were in operation, producing a maximum of 700 tons of molten pig iron a day. Then about the time of World War II steam turbines and turbine blowers came into general use, and iron smelters had available greater quantities of air. Now a new difficulty developed. To force greater amounts of air upward through many feet of furnace burden, the pressure had to be so high that at times the ore and fuel could not settle and were held up in the shaft or even blown out the top. To reduce the pressure, it became necessary to use more care in sizing and selecting the coke, ore, and limestone, and to refine the method by which they were charged into the top of the shaft. Resistance to gas flow, however, remained the factor that limited the amount of air that could be blown into a smelting furnace, and this, in turn, limited its ironmaking capacity. By the 1940s furnaces had reached about the size they are today—eighty to one hundred feet in height and twenty to thirty feet in diameter. When conditions were right—that is, when the ore was good and the coke was fast burning—one of these blast furnaces could make over 1,000 tons of iron a day.[4]

This was the situation in the 1940s when the University of Minnesota's Mines Experiment Station was seriously studying methods of processing taconite. It was hoped that some way could be found by which this low-grade material could be made commercially competitive with high-grade ores. Minnesota's famous, rich, high-grade iron ores were being rapidly depleted, but the rock called taconite was known to be enormously plentiful. It exists in vast quantities on the Mesabi

Range in Minnesota and elsewhere in the United States, in Canada, Russia, Manchuria, Australia, and other areas. While the name "taconite" originally referred only to the Mesabi deposits, it has by general usage come to mean rock of this type found anywhere in the world. The Russians, for example, have also adopted the term for the large deposits of taconite now being mined and processed in the Ural Mountains.

It was obvious to the author and his co-workers on the staff of the Mines Experiment Station that taconite processing would be expensive. Therefore we set out to make a superior final product, one that would smelt more rapidly and produce high-grade iron and steel more cheaply than other available ores. After much study and discussion, we decided that the taconite product best suited for smelting would be in the form of small spheres of high-grade concentrate. These uniform, spherical pellets would permit the greatest use of air and the best circulation of gases through the blast furnace burden. Our theory was that for optimum results, taconite pellets should be smelted unmixed with other iron ores, for irregularly shaped particles would reduce and alter the shapes of the voids between the balls, thus making a tighter furnace charge that would interfere with proper gas circulation.

While the experienced blast furnace operators with whom we discussed this program were not overly impressed by such theoretical considerations as microporosity, reducibility, and density, they were much interested in the possibilities of the constant day-after-day, year-after-year, unvarying structure and analysis of the pellet that could be produced. Naturally they did not care to express themselves definitely until they had available enough pellets for a thorough test in one of their big furnaces. But such a test would take 100,000 tons of pellets a month for several months, if our calculations were right—far more than could be produced in our laboratory.

And so it was not until 1960, after the world's first large taconite processing and pelletizing plant went into production at Silver Bay, Minnesota, that the blast furnace operators of two steel firms had the opportunity to experiment with, and finally test, the pellets in one of their big furnaces over a sufficiently long period of time. The results were startling, and a new record in pig-iron production was set.

Behind this achievement lies a long story—the story of taconite development in Minnesota. It is a story to which the author contributed a chapter, but which had its beginnings almost a century ago when two pioneers discovered this hard rock. In 1870 Christian Wieland of Beaver Bay, Minnesota, led Peter Mitchell of Ontonagon, Michigan, into northern Minnesota and showed him what Mitchell came to believe was "an iron mountain twelve miles long." Mitchell's mountain was one of the earliest discoveries of iron ore in place on the Mesabi Range, although it turned out to be taconite, a then valueless iron-bearing rock. We now know that this exceedingly hard rock can be enriched (or beneficiated) by mechanical processing. This book attempts to record the events leading to the first use of this rock and to describe the development of taconite processing in Minnesota. Peter Mitchell's mountain of iron was real after all. But it was a mountain of small iron pellets locked up and hidden in the old, old rocks of the Mesabi Range.

1 An Iron Mountain on the Mesabi

THE TACONITE STORY really begins in the 1860s in the frontier town of Ontonagon, Michigan, on the south shore of Lake Superior. Long before the Minnesota iron ranges were discovered, unusual deposits of metallic copper on Michigan's Upper Peninsula attracted the attention of travelers and explorers. Beginning in the 1840s copper mines were developed near Ontonagon, and it was a thriving city with saloons, hotels, and churches at a time when Duluth, Minnesota, and Superior, Wisconsin — at the western end of Lake Superior — were little more than trading posts. But the Ontonagon copper boom faded, and Michigan mining men began to look elsewhere for profitable mineral investments.[1]

It was natural that those who had not found their pot of gold in Michigan should look at the other end of the rainbow across Lake Superior in the young state of Minnesota. The north shore of Lake Superior there was opened to settlement in 1855 when Congress ratified the Treaty of La Pointe between the Chippewa Indians and the United States government. After that, prospectors began to drift into the region, and rumors of the presence of gold, silver, and copper filtered back across the lake.[2]

One of the men who seems to have directed the attention of Ontonagon's residents to the possibilities of iron ore in northern Minnesota was Christian Wieland. With his four brothers,

Christian in 1856 pioneered the settlement of Beaver Bay on Lake Superior's north shore. The Wielands were experienced woodsmen who ranged widely over the country between the lake and the Canadian border. In 1865 Christian guided Henry H. Eames, Minnesota state geologist, and his party to Vermilion Lake near Minnesota's northern border to investigate rumors of the occurrence of gold there. On the way, the group crossed the eastern end of the iron formation that later came to be known as the Mesabi Range, near what is now Birch Lake and the town of Babbitt. Christian is said to have pointed out to Eames the presence of iron ore in the rocks. The party, however, pushed on to Vermilion Lake, where the geologist found indications of gold. Rumors of his discoveries set off the so-called Vermilion gold rush late in 1865 and led to increased interest in the mineral potential of northern Minnesota.[3]

Christian Wieland seems to have been more interested in the iron ore he observed west of Birch Lake than in gold. He brought out samples of high-grade magnetic iron ore and exhibited them in Duluth and Ontonagon. The Wielands had built a sawmill near the mouth of the Beaver River, and they traded their lumber for supplies across Lake Superior in the Michigan city. Wieland's samples of iron ore created considerable interest there, and apparently led to the formation about 1869 of a loosely organized group, known as the Ontonagon Pool or Syndicate, whose purpose was to explore mineral areas in Minnesota and acquire land if the situation was as rosy as it was painted.[4]

No explicit records have been found to document just how or when the Ontonagon Syndicate was formed, but the pattern of similar groups is known. They were very informal organizations, and joining one was somewhat like joining a club. The members shared the expense of sending a competent explorer into an area to recommend what land should be secured. Each member then obtained as much or as little of the recommended property as he wished by whatever means he chose. Later a holding company was usually formed, and the land was transferred to it. Pool members then took stock in the firm in proportion to the amount and value of the land they had turned over to it.

We can imagine a group of men gathering in the bar of Ontonagon's Bigelow Hotel to discuss with Christian Wieland the great opportunities to be found across the lake in Minnesota. Apparently William Willard, an Ontonagon businessman, was the moving spirit of the pool. The other members were: James Mercer, Lewis M. Dickens, and Linus Stannard, all of whom were merchants or warehouse owners; Louis J. Longpre, a saloonkeeper; William W. Spalding, promoter; Peter Mitchell, prospector; and Samuel Mitchell, mine operator and capitalist. Later the pool was enlarged to include George C. Stone, Clinton Markell, Calvin P. Bailey, Joshua B. Culver, Daniel G. Cash, John C. Hunter, and Josiah D. Ensign, all of Duluth; John D. Howard of Ontonagon; Alexander Ramsey, a former governor of Minnesota then serving in the United States Senate, and perhaps others. It was reported that the Wieland brothers were to receive a one-fourth interest in any property acquired by the Ontonagon Pool in return for the information they could provide about the iron country.[5]

Before buying any real estate, however, the Ontonagon men wanted to check the information furnished them by Wieland. They therefore selected Peter Mitchell, a member of their group, to go into the iron country with Wieland and report to them on what he found there. Mitchell, who was forty-one years of age in 1870, was a big man with a heavy black beard and a great deal of determination. He had moved from Green County, Wisconsin, to Ontonagon in 1849. When he joined the Ontonagon Pool, he was apparently an experienced prospector. He was said to be something of an authority on geology and the identification of minerals, of which he accumulated an impressive collection of samples. He had the confidence of the Ontonagon businessmen, and his availability may have had a great deal to do with the organization of the pool.[6]

Setting out for Minnesota perhaps about 1870 or so, Mitchell probably went first to Beaver Bay, where he was outfitted and provided with packers and equipment at the Wielands' trading post. No account of his trip inland with Christian Wieland has been found, but the two men may have followed the same route that Christian had taken in 1865 when he guided the Eames party. If so, Mitchell and Wieland went to Greenwood Lake over a new road built by the Wielands, and from there to the Mesabi they used the old Indian trail along the Stoney River to Birch Lake.[7]

In this region, just south of Birch Lake and west of the Stoney River, the country is very rough and broken, with many rock exposures and great blocks and boulders of hard, gray-banded iron formation. West of the Dunka River the magnetic iron formation flattens out into a mile-wide stretch of comparatively smooth rock. In the exposures Mitchell probably recognized the narrow darker bands, sometimes an inch or two thick, as high-grade magnetite iron ore similar to the samples Wieland had secured there in 1865. If wide bands of this material could be located, the property would be very valuable. It is probable that, like other early prospectors on the Mesabi, Mitchell believed the high-grade ore would be found below the lean ore as it was on the Marquette Range, which was opened in Michigan in the 1850s. And so the search began.[8]

South of Birch Lake the rugged country of the Giant's Range, called "Missabay Height" by the Wielands, was heavily forested in 1870.[9] The hills there rise precipitously to a height of 300 or 400 feet above the adjacent country. Near Wieland's Missabay Height — close to the present town of Babbitt — these hills reach an elevation of slightly over 1,800 feet, which is 1,200 feet above Lake Superior. When approached from the north, the Giant's Range is an imposing granite barrier, but on the southern slope — where the Mesabi iron formation lies — the surface descends gently toward Lake Superior at a slight angle.

Mitchell and his party sank several test pits in the taconite rock near Babbitt. The one located in Section 20, Township 60 North, Range 12 West, is said to be "one of the very first pits ever put down" on the Mesabi Range.[10] Sinking pits in the extremely hard taconite rock was a formidable job for Mitchell and his crew; without mechanical equipment it would be a difficult undertaking today. With only the hand tools that could be packed in from Beaver Bay, it must have taken a crew of men several weeks to put down a six-foot test pit. Picks and shovels would have been of little use. The task required heavy sledges, hard drill steel, and chisels, as well as a blacksmith equipped to sharpen and temper them. Undoubtedly some blasting powder was also used.

Although we can only surmise why Mitchell chose certain locations and went to so much work to put down a number of these shallow pits, it seems probable that he hoped to find thick seams of high-grade ore similar to those in Michigan. From his experience, he would expect the good, hard ore, which could be mined by the underground methods customary at that time, to lie beneath the lean ore just above the underlying slates and granites. The soft, earthy ores, which lay near the surface and could be mined with shovels from great open pits, farther west on the Mesabi were still unknown. Near Birch Lake Mitchell apparently found ore to his liking—what he thought was "an iron mountain about twelve miles long and a mile and a half wide."[11]

Mitchell and Wieland left the Mesabi and visited the nearby Vermilion Range. There they viewed exposures of hard hematite, but Mitchell seemed singularly unimpressed. The Vermilion Lake area does not have the wide, flat exposures of iron formation which can be seen on the eastern Mesabi. Generally the Vermilion deposit is covered deeply with sand and gravel, and the surface is broken only here and there with exposed outcrops of hard hematite. What iron ore Mitchell could see there probably looked small and scattered compared to the magnetite near Birch Lake.[12]

Moreover, at the time of Mitchell's visit neither area had been surveyed. George R. Stuntz, an enthusiastic promoter of the Vermilion district, occupied the post of government surveyor at Duluth. This position made him an important person, for he classified the land he surveyed as swamp, farm, or mineral, and thus controlled not only the method by which it could be acquired but also the price. Mitchell's choice between the Vermilion and the Mesabi may well have been influenced by the fact that Stuntz was interested in the Vermilion. Wieland and Mitchell may have suspected that the Ontonagon Pool members would not be able to secure title to good mineral lands there except at prices they would consider exorbitant. If all went well, the two men probably expected to acquire the Birch Lake land, after it was surveyed, for as little as $1.25 an acre. Added to these reasons was Mitchell's conviction that he had found an iron mountain near Birch Lake.

Pool member Senator Ramsey was doubtless helpful in obtaining government authorization for an official survey of the eastern Mesabi and in having the sympathetic Christian Wieland appointed deputy surveyor. Wieland began his work on Township 60 North, Range 13 West, in February, 1872, and Ramsey kept track of his progress. The following September, while Ramsey was in St. Paul, the surveyor visited him before filing his report on Range 13. A month later Wieland was back in the state capital to file his notes on the adjoining Range 12, where he and Mitchell had put down their first test pit. Although Wieland did not file the second set of field notes until October 29, the surveyor again visited Ramsey on October 22. The following day the senator noted in his diary that he held two conferences with Wieland, "Willard," and Charles T. Brown, Minnesota's surveyor general. It seems probable that the Willard referred to was William Willard, the leader of the Ontonagon Pool, who may have journeyed to St. Paul to confer with these men and make certain the pool members secured the lands they wanted.[13]

After 1865 Minnesota land could be acquired in a number of ways, but the cheapest method was by homesteading. In theory, a person homesteaded a piece of property for the purpose of living on it and farming it. However, Congress so frequently changed the laws by which mineral lands could be secured that Minnesotans were never sure just what was legal and what was not. For example, a Congressional act in 1841 excluded mineral lands from the rights of pre-emption as well as from grants to railroads, universities, or schools. But in 1850 the attorney general ruled that lands containing iron ore were not mineral lands and therefore did not come under the 1841 act.[14]

As a result of the confusion caused by rapidly changing interpretations of the laws, by slow communication of these changes from Washington to Minnesota, and by outright disregard for legal requirements, there developed certain methods of acquiring land in the northern part of the state. While these violated the intent of the law, they were, nevertheless, practiced openly and condoned by the public, the government's land officers, and even the courts. Anyone wishing to acquire a piece of property could hire at Duluth a professional homesteader who built a crude house on the desired land and lived there for the prescribed length of time — or at least signed an oath that he had done so. The homesteader then received a deed to the property, which he immediately assigned to the person employing him and collected his fee.

As soon as Wieland's survey of the eastern Mesabi was completed, the Ontonagon Pool members put homesteaders on the land. William Willard, acting either for himself or as an agent for others, located some twenty-four homesteaders in Township 60 North, Range 13 West, in 1873 and 1874, and later secured title to 3,840 acres. Other pool members acquired land in Townships 59 and 60 North, Ranges 12, 13, and 14 West, comprising all the iron land on the eastern Mesabi and some beyond the limits of the mineral area. In all, at least 9,000 acres came into the ownership of the Ontonagon businessmen.[15] Although they did not know it at the time, these men had not acquired the high-grade ore lands they sought; instead they had become the first owners of an enormous supply of taconite.

Unaware of this fact, the pool members in 1874 incorporated the Duluth and Iron Range Railroad Company in the belief that a railroad was needed to exploit their property. Listed as incorporators were Peter Mitchell, nine other members of the pool, and three additional Duluth businessmen — Luther Mendenhall, Benjamin S. Russell, and William R. Stone. William Spalding was elected president and Calvin Bailey, secretary and treasurer. In 1875 the firm was successful in a maneuver that was later to have important results. It secured a franchise and a state land grant of ten square miles for each mile of track built and put into operation between Duluth and the Mesabi Range.[16]

No immediate progress was made in constructing the railroad, but in 1882 the Ontonagon Pool members incorporated a new firm — the Mesaba Iron Company — to develop their mining property. Ramsey was elected president of the board of directors, and Spalding acted as secretary and treasurer; both men served in these offices until their deaths just after the turn of the century. The articles of incorporation allowed the company wide leeway to carry on the mining, smelting, and manufacturing of iron, steel, and other metals, and the buying, selling, and leasing of mineral and other lands. The capital stock was set at $3,000,000, divided into 100,000 shares, of which half were to be distributed to the stockholders in exchange for the land they would convey to the company.[17]

Although most of the Mesaba Iron Company's directors and stockholders had been members of the Ontonagon Pool, the roster included some new names. William Harris, an Ontonagon mining captain, and William D. Williams, a lawyer of Ontonagon, Marquette, and later Duluth, joined the group. Other names that we would expect to find on the list are missing. William Willard had died suddenly in 1874. Since he had been the directing head of the Ontonagon Pool, his loss, coupled with the Panic of 1873, undoubtedly had a very disturbing effect on the organization and its newly incorporated railroad company. Christian Wieland died in 1880, but among the Mesaba Iron Company's incorporators was his nephew, Henry P. Wieland. He seems to have been the agent or representative for other members of the Wieland family, many of whom are known to have later owned small holdings of stock in the company. But what happened to Samuel Mitchell and L. M. Dickens? Perhaps they, too, were represented by others, or perhaps they did not choose to join the pool members in forming the company.[18]

The name of George Stone is also missing from the list of Mesaba Iron Company stockholders. In 1869 Stone had arrived in Duluth from St. Louis, Missouri, as an assistant to George B. Sargent, Jay Cooke's agent, and he soon became the working head of Duluth's first bank. A smooth-talking man who inspired confidence, Stone was elected treasurer of the city of Duluth in 1870 and entered the Minnesota legislature in 1876. He was a friendly, optimistic promoter who quickly became interested in many Duluth business ventures. In 1875 he traveled east in search of far-sighted men with sufficient money and interest to support a mining development on the pool's Mesabi Range lands.[19]

In his travels Stone called on wealthy men in Chicago, Detroit, and Cleveland, but it was not until he met Charlemagne Tower of Pottsville, Pennsylvania, that he received an encouraging reception. Tower had heard of Minnesota's iron ore, and Stone's high-grade samples from Mitchell's mountain on the Mesabi interested him greatly. In 1875 he sent Albert H. Chester, a mineralogist and chemist at Hamilton College, Clinton, New York, to investigate. Chester traveled by canoe from Duluth to Vermilion Lake, where he looked over the "hematite exposures." Then, from a base camp near the later town of Mesaba, he made a close study of the Birch Lake area of the eastern Mesabi. He discovered Peter Mitchell's first test pit, had it cleaned out and deepened, and the iron content on one side of the pit analyzed. He found the total thickness of ore to be about 3.5 feet out of a total of 11.2 feet; a sample made from all the best layers assayed 44.10 per cent iron. Since Mitchell's samples had yielded an analysis of over 50 per cent iron, it was apparent that they had been hand-picked.[20]

Chester also found what he thought was Mitchell's mountain of ore, but it, too, proved disappointing. Although selected narrow bands were high-grade ore, the average was quite poor. The professor wrote: "The whole amount of good ore here shown is less than two feet, the best being at the very top of the cliff and showing 62.37 per cent" iron. He admitted, however, that while Mitchell's mountain was only about twenty-five feet high, "at the top, the whole surface for some distance is . . . smooth and polished by glacial action, and as one walks over it he seems to be walking on solid iron."[21]

After collecting a large number of samples over the fifteen or twenty miles between his camp and Birch Lake, Chester arrived at the conclusion, which he reported to Tower, that the Mesabi was not nearly as important as the Vermilion Range. Of course Chester's mistake was in looking to the east instead of to the west. He thought he had been studying the entire Mesabi, but he had covered only the eastern fifteen-mile extension of that large and then unknown range which stretched westward a hundred miles to the Mississippi River. Tower then wrote Stone in Duluth that he did not care to invest in the Mesabi — a report that very much distressed Stone. But five years later, after Chester made a second trip to the Vermilion, Stone was employed by the Pennsylvania financier — this time to buy up "an empire of iron land" on the Vermilion Range.[22]

By the end of 1882 Tower had acquired over 20,000 acres in northeastern Minnesota. On December 1, 1882, he incorporated the Minnesota Iron Company and transferred to it all his holdings. Stone became one of Tower's business partners. Through him the Minnesota Iron Company soon took over the valuable railroad franchise that the old Ontonagon Pool members had obtained in the hope of developing their Mesabi lands. Tower needed the Duluth and Iron Range Railroad to reach his Vermilion property, but the original legislative act of 1875 specified that the road was to have one terminus at Duluth and the other on the Mesabi Range. Therefore, in 1883, the financier sent Stone to St. Paul, where he was successful in getting the legislature to amend the act and relocate the northern terminus on the east side of Vermilion Lake. Stone also succeeded in retaining for Tower the Duluth and Iron Range Railroad's swamp land grant of ten square miles for each mile of road actually built.[23]

After several surveys were made, Tower decided to build the railroad from Vermilion Lake to Agate Bay (now Two Harbors) on Lake Superior, crossing the Mesabi Range at the town of Mesaba. In 1883 a contract was let to construct the road, and on July 31, 1884, the line was completed. The last rail had scarcely been laid at Two Harbors when Minnesota's first trainload of iron ore arrived after a run of sixty-eight miles from Tower's newly developed Soudan Mine on the Vermilion Range near the town that bears his name. Two years later the railroad was extended to Duluth, and Tower secured from the state his land grant of over 600,000 acres.[24]

With the completion of the railroad, Minnesota quickly became important to the nation's growing iron and steel industry. It was immediately demonstrated that hard, high-grade Vermilion ore could be delivered to Cleveland and other ports on the lower Great Lakes in competition with ores from the older ranges on the south side of Lake Superior. Moreover, the new railroad intersected the Mesabi Range and thus gave easy access to that as yet undeveloped iron ore district. Exploration mushroomed. In a few years the town of Mesaba became the starting point for prospectors working both to the east and the west. Roads of sorts were built, connecting the communities which began to grow up about the mines.[25]

The Mesaba Iron Company, incorporated by the Ontonagon Pool in 1882 after Tower began to buy up land on the Vermilion, did not develop its property; instead the firm tried to sell its holdings on the Mesabi at a profit. The company granted options to buy the land to various promoters, none of whom made any valuable discoveries. While the drilling and prospecting appear to have given the company little encouragement, the presence of Tower's railroad some five miles from the property created some interest. The firm upped its asking price in 1892 after the Merritt brothers began to develop their famous Mountain Iron Mine, which brought the Mesabi Range to the attention of the nation. Thirteen years later, on December 13, 1905, the stockholders voted to sell all the real property of the Mesaba Iron Company "to George A. St. Clair, of Duluth, Minn., for . . . $100,000," less a commission of $10,000 which was to be paid to St. Clair. The land sold consisted of 8,840 acres in Township 60 North, Ranges 12 and 13 West, and included Peter Mitchell's old test pits. For this property the company received more than ten dollars an acre — a good price at a time when much government land in the state was still being sold for a dollar or two an acre.[26]

Two other men were associated with St. Clair in the purchase of these lands. They were Samuel Mitchell, one of the original incorporators of the Ontonagon Pool, and John G. Williams, a Duluth lawyer. Both St. Clair and Mitchell were experienced mining men who owned and operated properties in Michigan. In 1905 they worked together in the early development of the Athens Mine at Negaunee, and it was probably this association, plus Mitchell's relations with the Ontonagon Pool, that led to the sale of the Mesaba company's lands. Moreover, St. Clair was interested in the Spring or Siphon Mine, the most easterly hematite mine on the Mesabi, which adjoined the Mesaba Iron Company's property. The third man, Williams, had settled at Duluth in 1884. He served as attorney for many of the pioneers in Minnesota's iron ore industry, and it was in this capacity that he came to know St. Clair and Mitchell.[27]

These men in August, 1906, organized two companies to hold and develop the property. At that time Minnesota had a law limiting to 5,000 acres the amount of land any firm could own in the state. To take care of their holdings, which totaled nearly 9,000 acres, the three men formed the East Mesaba Iron Company and the Dunka River Iron Company. In each, St. Clair owned a six-tenths interest, Mitchell a three-tenths interest, and Williams a one-tenth interest. The firms were incorporated on August 29, 1906, with a capitalization of five hundred shares each. The par value of a share was set at $100.[28]

Two years later Samuel Mitchell died. With his death, the last of the Ontonagon Pool members became disassociated with the Mesabi property brought to their attention by Christian Wieland in the 1860s, and a new chapter in the history of this land began. In retrospect, we can view the Ontonagon Pool as a group of far-sighted pioneers with great plans but without the experience or financial insight to carry them out. At one time, the whole Vermilion Range and the Duluth and Iron Range Railroad were within their reach, but they could not close their fingers. Instead, probably misled by Peter Mitchell's report of a mountain of iron, they chose land on the eastern Mesabi which was not to become useful for many years. Expecting to discover hard, high-grade ore, they explored only the small eastern tip of the mighty Mesabi Range and found a vast field of taconite. Had they looked farther to the west they would have found, embedded here and there in the taconite, the fabulously rich mines of soft, red hematite that have fed the steel furnaces of the nation for over half a century.

2 The Mines Experiment Station Encounters Taconite

MY INTRODUCTION to taconite occurred five years after Samuel Mitchell's death. In a roundabout way, John G. Williams, one of the men who had purchased the lands of the old Mesaba Iron Company, was responsible for my first contact with this hard rock. He was in 1913 a regent of the University of Minnesota. In those days, the university at Minneapolis was a comparatively small institution with an enrollment of about four thousand students, and after a meeting, the regents, and sometimes even the president, went about the campus visiting various departments and calling on friends. Regents Williams and Fred B. Snyder, both of whom were interested in mining, often visited Dean William R. Appleby at the School of Mines.

On one of these occasions early in 1913, the dean called me in, as a new instructor in mathematics, to introduce me to the regents. At that meeting, Williams told us about a great area of mineral land of which he was part owner. It was, he said, on the eastern Mesabi Range between the town of Mesaba and Birch Lake, and it was entirely composed of taconite. But locked up in the rock, Williams maintained, was more iron than all the known deposits of high-grade ore in the state put together. It just remained for someone to find out how to extract this iron economically.

We listened politely, if somewhat skeptically, and Dean Appleby, always the diplomat, suggested that Williams send a sample of this rock to the newly formed Mines Experiment Station, which the dean had organized for the purpose of studying the minerals of the state. In the spring of 1913 a sack of rough greenish-gray rock arrived. Although I was not yet formally a member of the Mines Experiment Station staff, Dean Appleby assigned to me the job of studying the sample. This was the beginning of my contact with taconite and the beginning of the Mines Experiment Station's work with it which was to stretch out over forty years.

For the first twenty-two years of this period Dean Appleby was the responsible head of both the School of Mines and the Mines Experiment Station. He had been appointed to the university's staff in 1891 to organize a mining and metallurgical department. At that time there was still much interest in gold in northern Minnesota, and his first project had been to install an assay laboratory in Pillsbury Hall, where students could be instructed in the methods of determining positively and accurately the gold and silver content of samples of rock collected from the Vermilion district and other sources. In 1892 Appleby secured $5,000 in contributions from local business people and created the Ore Testing Works, which in 1911 became the Mines Experiment Station.[1]

When we began the study of Regent Williams' taconite, it was not only a bit of "apple polishing," but also the kind of undertaking Dean Appleby considered an essential part of a great university. He was a firm believer in the practical application of academic theory. What good was academic training if it could not be put to some practical use? This idea did not make him popular with some members of the university staff, but he resisted all attempts to change his educational program. After Lotus D. Coffman became president in 1920, he and Appleby had an almost continuous battle. Guy Stanton Ford, dean of the Graduate School and later university president, once said, "When Appleby retires we are going to annex the School of Mines to the University of Minnesota."[2]

Under Dean Appleby's tutelage the Mines Experiment Station flourished and work on taconite began. When I became a formal member of the station's staff in 1914, it was housed in a small wood and brick building formerly occupied by the Ore Testing Works on the bank of the Mississippi River. A fire had gutted the School of Mines building early in 1913, and a new structure was being erected. This meant that the Ore Testing Works would no longer be needed for classroom demonstrations, and it was slightly remodeled and modernized to take care of the expanding needs of the Mines Experiment Station. We had our own machine shop, analytical laboratory, and a helper or two, and in those days we had to make a large part of our equipment. In the first place, we did not have much of a budget to draw on, and in the second place, very little suitable equipment was then available anywhere for processing taconite.

The sample Williams sent in was the first taconite we had ever examined. It was a hard rock, mottled in appearance, with a few narrow bands of darker material passing through it. Microscopic examination showed that the iron existed as small black particles of magnetite embedded in siliceous rock. When powdered, much of the rock was a pale greenish color and had a glassy appearance, although occasionally small, gray, opaque particles of quartz could be observed. After breaking the sample into small pieces, I found that most of them could be picked up readily with a hand magnet. The pieces from the dark bands were attracted more firmly than those from the light areas, but a powerful magnet would pick up practically all of them. I found that the dark, strongly magnetic particles contained from 40 to 60 per cent iron, while the light-colored, less magnetic ones contained only 10 to 20 per cent. Occasionally a little piece could be selected that would stick weakly to the magnet, if at all, and this would assay only 5 to 10 per cent iron. The concentration of black magnetic particles was higher in the dark bands, but even in the light areas small pieces of magnetite were numerous.

From this preliminary examination it was obvious that only a fraction of the iron was concentrated in the dark bands, and even these contained many particles of greenish-gray rock. Pure magnetite particles free from inclusions of rock could be observed under the microscope, but they were very small. To release any large percentage of the magnetite from the attached and included particles of rock would obviously require crushing to a very fine size. Areas of quite pure magnetite as large as a sixteenth of an inch square could be seen, but they were rare. To add to the difficulty, the structure was much finer in some samples than in others. On the average, however, it seemed at the time that the rock would have to be crushed to a powder having no particles substantially larger than a hundredth of an inch in diameter in order to completely free any large percentage of the magnetite.[3]

It was no easy task with the equipment then available at the university to powder the hard rock that fine. Moreover, after we had powdered it, we found it difficult to remove the particles of good magnetite even with a strong magnet. After a piece of taconite has been crushed fine, the powder can be divided into three products: concentrate, which is high enough in iron content to be of value; tailings, which are low enough in iron content to be discarded; and middlings, which are an intermediate product too good to be discarded and too lean to be considered concentrate. (See Figure 8.) I was not able to make a concentrate containing as much as 50 per cent iron, but I was sufficiently interested in the work to look up some information on magnetic separation, about which I knew little.

I learned that wet concentration of similar fine magnetic ore had been used to a limited extent at plants in Pennsylvania, Canada, and Norway. In some notes made at the time, I remarked that the ore was "very hard and extremely dusty and some wet concentration method should be used." Later I made tests with a hand magnet under water and immediately secured much better results. The first wet test was made in October, 1914, on pulverized crude ore assaying 29.11 per cent iron. Laboriously made in water with a hand magnet and many washings, it produced a concentrate assaying 68.88 per cent iron and 3 per cent silica. When the results of the test were reported to Regent Williams, he was much encouraged, and thereafter both he and George St. Clair came to the laboratory every week or two to see how the work was progressing.[4]

Williams suggested that we try to devise equipment that would more easily wash the silica out of the powdered ore. I was certainly in favor of that! As a result of his suggestion, I secured funds from the university and developed a rather cumbersome wet magnetic separator.[5] It consisted of a large C-shaped electromagnet between the poles of which a glass tube about one inch in diameter and two feet long was supported at an angle of about forty-five degrees. Water flowed into the top and out of the bottom of the tube through control valves adjusted to keep the tube full of water at all times. When finely pulverized taconite was placed in the tube, it immediately collected between the poles of the magnet. Moving the tube up and down a few inches while rotating it, freed the silica, and the water then washed it downward and out of the tube, leaving the good magnetic ore collected at the poles of the magnet. This device became standard equipment for testing small samples of magnetic ore. We called it a "Magnetic Tube Separator," and it is still in use, with some refinements, in most laboratories working on taconite or other magnetic ores.[6]

The development of this apparatus simplified the concentration of pulverized samples weighing an ounce or two. With most of the rock that Williams sent to the university, high-grade ore concentrate assaying over 60 per cent iron could be made. Ore shipments from the Lake Superior district at the time (1914) averaged 51.34 per cent iron; thus Williams and his associates were greatly encouraged.[7] All that was required, they felt, was to crush and grind the taconite to a fine powder and then concentrate it magnetically in a machine which would produce the same washing action on a large scale that the tube did on a small scale.

I tried to develop such a machine and was moderately successful. In 1915 we built what is known as a magnetic log washer in our shop. It had two parallel troughs five inches wide and twenty-four inches long which were lined with copper. Beneath the copper sheet was an electromagnet, with one pole in the center between the troughs and one on each side along their full length. In the troughs and extending the length of the machine, two shafts with small blades or paddles were mounted so they could rotate. The machine was set on a slight incline with a dam in the trough at the lower end. A spray of water came in at the upper end, flowed down the troughs, and overflowed the dam at the lower end. When fine magnetic ore was fed into the machine, magnetite collected on the bottom of the troughs near the magnets and was moved upward and out of the water by the action of the rotating paddles. The nonmagnetic material was carried by the flow of water to the lower end of the machine and flowed out over the dam. The device was, in principle, very much like the standard log washers then being used on the sandy ores of the western Mesabi, but with an electromagnet added.[8]

Our small laboratory machine could operate continuously on samples of a few pounds, producing results very similar to those we secured with the tube on a few ounces. The first test on this machine, made on January 20, 1915, using eastern Mesabi taconite crushed to 40 mesh, produced concentrate assaying 40.83 per cent iron. On January 28, working on 150 mesh ore, we produced concentrate assaying 60.44 per cent iron and tailings assaying 4.76 per cent iron.[9]

When Williams saw the machine in operation on February 8, he was very pleased because the tests at the Mines Experiment Station tended to confirm other experimental work in the processing of Mesabi taconite. By the time the station received its first taconite rock in 1913, there was already considerable interest in the potentialities of the land held by Williams and St. Clair. The story is somewhat complicated, but the key figure is Dwight E. Woodbridge, a Duluth mining engineer.

In 1905 Woodbridge had been hired by Hayden, Stone and Company, a large banking and brokerage firm of Boston and New York, to investigate a new copper property in Utah which was being developed by Daniel C. Jackling. As a result of Woodbridge's favorable report, Hayden, Stone acquired a large interest in Jackling's Utah Copper Company. The investment proved to be very profitable and led to the development of several other Western low-grade copper properties.[10]

Seven years later Woodbridge was hired by Williams and St. Clair to make a similar investigation of their taconite lands and to recommend how they could best be exploited. Woodbridge's conclusions, though no doubt disappointing to his employers at the time, have turned out to be remarkably accurate. After diamond drilling, test pitting, and sampling, he reported that the Birch Lake property contained no mineable deposits of high-grade ore, but that the whole area was one great mine of taconite. Comparatively complex processing, Woodbridge told Williams, would be required to make good ore from the taconite. Simple methods—such as the selective mining of high-grade bands on the Vermilion and on the older Michigan ranges, or the washing process then being developed farther west on the Mesabi—could not be used with this low-grade rock. He added, however, that other mines in the United States, Canada, and Norway had utilized magnetic concentration on ores of a somewhat similar nature.

In 1913 Woodbridge called the attention of Hayden, Stone and Company to the large, low-grade iron ore deposit on the eastern Mesabi. Galen Stone, one of the firm's senior partners, then arranged a conference between Woodbridge and Daniel Jackling, with the idea that the methods Jackling had so successfully developed for low-grade copper ore might be applied to low-grade iron. Several more meetings followed, and Woodbridge was asked to provide additional information on the concentration methods that would be required and on the costs involved.[11]

He arranged to have two carloads of taconite (80,800 pounds) from the Williams-St. Clair lands shipped to the magnetic concentration plant of Moose Mountain, Limited, near Sellwood, Ontario. There the taconite was crushed and concentrated under the direction of the plant's superintendent, Fred A. Jordan. A former instructor in the Michigan College of Mines at Houghton, Jordan was later to play an important part in the development of east Mesabi taconite.

In those days, Moose Mountain was a small plant struggling to become economically successful. It was experimental in nature, having dry cobbing, fine grinding, and wet magnetic concentration equipment, which could be adapted to the testing of small shipments of ore. Cobbing is the term applied to the separation of the ore after it has been crushed into coarse particles, usually smaller than three inches and larger than a half inch. Dry cobbing was done on a machine consisting of a short conveyer belt and a magnetic head pulley. The dry ore was fed onto the conveyer belt in a thin layer; as it passed over the magnetic pulley, the weakly magnetic pieces of taconite fell off and the strongly magnetic pieces were carried farther around the pulley before they, too, fell off. The amount of each product depended upon the nature of the ore, and to some extent the quality could be controlled by adjusting the strength of the magnets.

In October, 1914, Woodbridge reported the results of the work at Moose Mountain in a letter addressed to N. Bruce MacKelvie of Hayden, Stone. The samples had been taken from two locations in Section 19, Township 60 North, Range 12 West, near Peter Mitchell's second test pit. At Moose Mountain the taconite from the first spot assayed 33.84 per cent total iron and from the second, 34.60 per cent. However, as Woodbridge explained, the total iron assays had little meaning to concentrating engineers because some of the iron was not what he called "free" or "soluble," that is, in the form of magnetite which could be recovered by magnetic concentration. The free iron in the two samples was reported as 28.15 per cent and 29.12 per cent.[12]

At first the Moose Mountain plant crushed the rock to a coarse size—somewhere between one inch and one-quarter inch—and then removed magnetically the pieces of high-grade ore that occurred as dark bands in the taconite. Woodbridge found this unsatisfactory, and he told MacKelvie that "these results proved the impracticability of taking off a high grade product." In other words, the dark bands were not rich, thick, or numerous enough to be processed economically after coarse crushing.

Woodbridge then tried grinding the samples finer. He learned that by grinding to 60 mesh he was able to recover almost 90 per cent of the iron as a concentrate assaying 57.45 per cent. He told MacKelvie that by grinding still finer—to 150 mesh—a concentrate assaying over 60 per cent iron could be made. He noted that at the coarser sizes some 25 per cent of the taconite could be rejected magnetically as rock too poor for further treatment, thus reducing the amount requiring fine grinding.

The engineer outlined at some length the possibilities of selling the coarse, rejected rock for construction purposes, suggesting that the price which could be obtained for this by-product "would more than meet the cost of mining and separating it." He estimated that a plant for processing the taconite could be built for not more than $3.00 per ton of annual concentrate capacity. (Modern taconite plants cost over ten times this amount.) Optimistically he calculated that 60 per cent iron concentrate could be produced for $2.58 a ton, if the coarse rock were sold as concrete aggregate at 60 cents a ton; and that the concentrate would have a market value at the mine of $4.11 per ton, indicating a profit of $1.53 a ton. Woodbridge concluded his report to MacKelvie by saying, "You have a potential asset in this proposition that is at least as great as anything that can be pointed to, for the market is not only tremendous but the quantity of material to be treated is vast."

Although we now know that his estimates were overly optimistic and that some of his arithmetic was wrong, Woodbridge for the first time brought Peter Mitchell's mountain on the east Mesabi into proper focus and channeled thinking about it in the right direction. He accurately depicted a large deposit from which no high-grade ore could be obtained without fine grinding and elaborate and expensive processing.

Apparently Woodbridge's information interested Jackling, for he then sent Walter G. Swart of New York to investigate and report to him on the economic possibilities of developing low-grade iron ore in Minnesota, as he had developed low-grade copper in the Southwest.

Walter Swart was forty-seven years old when he encountered Minnesota taconite. A graduate of the University of Denver and the Colorado School of Mines, he had known Jackling since the two men had operated assay offices in Cripple Creek, Colorado, in 1900. Swart was a scholar and an optimistic dreamer rather than a rough, hard-boiled mining engineer. He was a kind, friendly person, and people in the Duluth and Babbitt areas will remember him as a rather heavy man with a smiling face and a bald head with just a fringe of hair.

Swart appeared in Minnesota in the spring of 1915 and visited the eastern Mesabi in the company of Woodbridge, Williams, St. Clair, and Horace V. Winchell, the son of Minnesota's former state geologist. On June 16, 1915, the group came to the Mines Experiment Station to see the ore being crushed and concentrated in the tube machine and in the newly devised magnetic log washer. Later Swart sent Jackling a very complete and glowing report on the whole taconite project, recommending that more diamond drilling be done on the east Mesabi property and that a complete survey be prepared by competent geologists. If such investigations showed the existence of the required tonnage of taconite, he suggested that a pilot plant be built and equipped with commercial machinery to demonstrate the feasibility of treating this rock to produce a 60 per cent iron concentrate. After that, he said, consideration could be given to constructing a large commercial plant near the property. Swart concluded: "There is no doubt in my mind that this is a game worth playing and that it can be played to a successful finish. It seems now only a question of men, money, methods, and details."[13]

Jackling decided to follow Swart's recommendations. To direct and finance the investigation, he joined with Hayden, Stone and Company in organizing the Mesabi Syndicate in 1915. In addition to Jackling, Charles Hayden, and Galen Stone, the syndicate was composed of Seeley W. Mudd, Bernard M. Baruch, Percy A. Rockefeller, Louis S. Cates, and John D. Ryan—all of whom had been associated in the Utah copper enterprise—as well as Ambrose Monell of International Nickel, William E. Corey of the Midvale Steel and Ordnance Company, Horace Winchell, Swart, Woodbridge, and others. Swart was to be the syndicate's resident Minnesota manager, and the others were to subscribe funds as needed. It is notable that the only members of the group from the iron and steel industry were Corey and Rockefeller, both of whom were directors of Midvale.

The syndicate then acquired control of the taconite lands held by Williams and St. Clair. On December 3, 1915, the Dunka-Mesaba Security Company was formed to take over the stock of the Dunka River and the East Mesaba Iron companies. The new firm then leased the property to Claude W. Peters, a member of Hayden, Stone and Company's organization. He, in turn, gave the syndicate the right to drill and explore the property and an option for a lease. Acting for the syndicate, he also secured additional adjacent land for commercial expansion.[14]

In 1916 Swart acquired an old lumber mill building at the foot of Thirty-ninth Avenue West in Duluth for use as a test plant. He also opened an office in the Sellwood Building there. Then he employed Fred Jordan, the former superintendent of the Moose Mountain plant, as his first assistant, and Theodore B. Counselman, a mining engineer and a graduate of Columbia University, who had been working on a low-grade copper property at Morenci, Arizona, as his second.

Swart also made arrangements for the Mines Experiment Station to carry out more extensive tests and experiments on the magnetic ore from the syndicate's taconite land. In January, 1916, Fred Jordan arrived at the station to act as Swart's representative in the tests to be conducted there. In May, Jordan's presence was required at Duluth, and Counselman replaced him at the station. Ted made a great many large-scale concentration tests on the syndicate's magnetic ore in an effort to work out the best operating conditions for the magnetic log washer we had devised. The annual report of the station for that year shows that a sample weighing 100,000 pounds was sent to us by Swart and that 2,500 hours were consumed in tests and experiments on it.[15]

In June, 1916, I took a leave of absence from the university and went to Duluth. Jordan, Counselman, and I constituted the syndicate's technical staff, and George H. Wallas was placed in charge of operations at the mine. James R. Mitten was employed as a chemist, and George H. Cregor was the sample room foreman. Later August Palo, who began work as an electrician, became foreman of the test plant and the laboratory.

The syndicate's activities on the east Mesabi got under way

with the establishment of a camp in Section 18, Township 60 North, Range 13 West, early in the summer of 1916. A logging railroad spur, built six years earlier, ran from Mesaba Station (about half a mile from the Mesaba townsite) east to the Dunka River, skirting the southern edge of the iron formation. Some old lumber camp buildings were still standing, although badly in need of repair, near a place on the railroad called Sulphur Siding, not far from Peter Mitchell's first test pit. They stood on high ground on the west bank of a small creek at the edge of a large swamp. The first job was to repair and equip these buildings as headquarters for crews that were to do the surveying, drilling, sampling, and the geological work, and later for a crew that would mine the taconite for the Duluth experimental plant.

The Sulphur Camp, as it was called, consisted of two large and two small log buildings. The two larger ones, each perhaps thirty feet by sixty feet in size, were fitted out as bunkhouse and cookhouse. One of the smaller buildings became the office, and the other served as the engineers' shack. The camp was quite comfortable and at times housed twenty or thirty men. It was, however, a remote and isolated place in the middle of a great burned-over area, where the blackened trunks of dead trees rose starkly from a tangled growth of brush. The only means of contact with the outside world was a gasoline speeder by which one could travel over the sixteen miles of railroad to Mesaba Station. From there one could get on a Duluth and Iron Range passenger train to Two Harbors and Duluth. Occasionally a locomotive came into Sulphur Siding, pushing a railroad car or two containing supplies and equipment. There was no passable road to the camp, although the old Syndicate Trail, originally cut out for Professor Chester's party in 1875, was still plainly visible in 1916.

The surveying crew was under the direction of Anthony F. Benson of Eveleth, who was later to become St. Louis County inspector of mines. During the summer of 1916 his men cleared out section lines and trails to provide access routes for the geological crew, surveyed the property, and prepared a complete contour map of the area. It was a slow job because the lines had to be chopped through head-high piles of half-burned "down timber." After lumber companies had cut all the good timber, a fire passed through in 1910 killing the small trees. These had since blown down, and they formed a tangle of brush and logs that made the country almost impassable, except where trails or roadways had been cut. Great roots of large, dead trees spread out in all directions on top of the bare, flat taconite. It was desolate country in 1916, and as nearly useless as can be imagined.

Because much more specific information was needed, Swart arranged with Dr. William H. Emmons, director of the Minnesota Geological Survey, to have his men work on the geology of the district. Before 1917 only a few diamond drill holes had ever been put down in the area around Sulphur. Peter Mitchell and others had an inkling of the general geological picture, but until Edmund J. Longyear brought his diamond drill to Mesaba on Tower's new railroad in 1890, there had been many opinions but little agreement because of the lack of factual information. Longyear sank the first diamond drill hole on the Mesabi Range about a mile and a half southeast of Mesaba Station. It was 1,293 feet deep and went completely through the iron-bearing rock, which we now call taconite but which Longyear called "magnetic quartzite." This was exactly the same material that Mitchell had encountered at the surface in his explorations thirteen miles to the east. Longyear's drill core showed no hematite but many bands of rich magnetite, none of which were thick enough to be mined separately from the rock in which they were embedded. He then moved his drill eastward over the Syndicate Trail to a location just south of Iron Lake, and there early in 1891 he drilled three more holes completely through the magnetic iron formation. No wide bands of good ore were encountered, however, and no hematite was found either by Longyear or by Woodbridge or Swart in their later drilling.[16]

Professors Frank F. Grout and Thomas M. Broderick of the Minnesota Geological Survey spent the whole summer of 1917 at Sulphur mapping the outcrops and studying the taconite formation. Three drill rigs were brought in, and geological drilling was done under the direction of these men. As a result of this exploration in 1916 and 1917, the area was mapped accurately and contoured by Benson and his crew, and a fine geological report was prepared by Grout and Broderick.[17] The work of these men not only gave the Mesabi Syndicate an accurate description of its taconite property, but also contributed materially to geological knowledge of the whole Mesabi Range.

The geological picture that has gradually developed from the work of Grout, Broderick, and others, is of a great sheet of iron-bearing sediments deposited on the bottom of a large body of water called the Animikie Sea.[18] Millions of years ago this sea occupied the whole western portion of the Great Lakes area. After the iron-bearing sediments were deposited in the Minnesota, Michigan, and Wisconsin iron regions, the boundaries of the sea were altered by the movement of the earth's crust, and the present basin of Lake Superior was formed by later earth movements. Some geologists believe that the iron-bearing rock of the Mesabi and Gogebic ranges is a continuous sheet dipping under Lake Superior and extending from the Mesabi on the north to the Gogebic on the south. This leads one to theorize that the amount of iron existing, in one form or another, in the Lake Superior district is enormous. (See Figure 1.)

It should be understood that this simple description does not span the known geologic time period of the region. At one time active volcanoes spewed out lava in dozens of great flows which are plainly evident along the north shore of Lake Superior, on Keweenaw Point on the south shore, and in the region between the lake and the Canadian border. Also, huge glaciers — sheets of ice thousands of feet thick — moved over this country several times, and their action scoured out the Lake Superior basin and left it as we find it today.

Some of the early explorers for iron ore guessed that the deposits were continuous under the lake. But what they did not understand was that the sediments had been laid down under different conditions on opposite sides of the lake. Although they had hardened into rocks of similar types, sediments deposited as far apart as the Mesabi and Gogebic ranges could be — and actually were — very different in their thickness and nature. The geological processes that altered the iron-bearing sediments to pro-

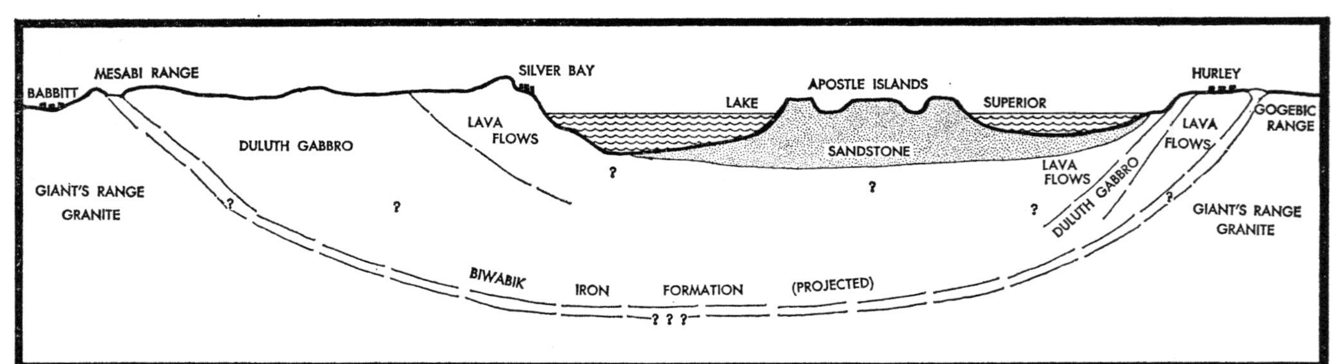

FIGURE 1. *Projected cross section under Lake Superior. Scale distorted. After Schwartz and Thiel.*

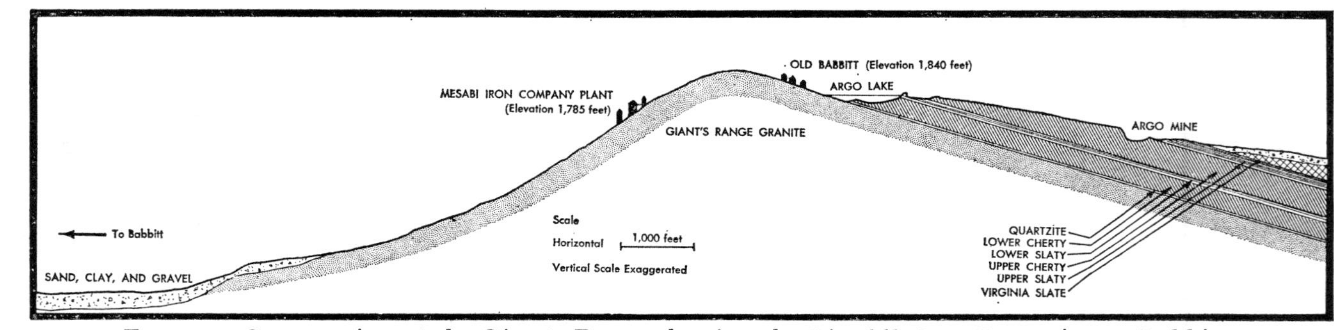

FIGURE 2. *Cross section of the Giant's Range showing the Biwabik Iron Formation at Babbitt.*

PETER MITCHELL, *prospecting for a group of Michigan businessmen, found what he thought was a mountain of rich ore on Minnesota's eastern Mesabi Range about 1870. It turned out to be taconite.*

THIS PIECE *of taconite shows the narrow, dark bands of high-grade ore which encouraged early prospectors like Peter Mitchell.*

THE TOP OF PETER MITCHELL'S MOUNTAIN *is covered with a smooth, two-inch layer of high-grade taconite, polished by glacial action. Mineralogist Albert H. Chester analyzed the ore below this layer in 1875 and found it disappointing. The author is shown pointing to one of the many glacial scratches.*

JOHN G. WILLIAMS *(left), Samuel Mitchell, and George A. St. Clair (right) in 1905 purchased the taconite land discovered by Peter Mitchell. It was Williams who sent the first sample of taconite to the University of Minnesota's Mines Experiment Station in 1913. Courtesy Oglebay Norton and Company.*

WALTER G. SWART *directed the work on taconite at Duluth and Babbitt from 1916 to 1924. This photograph was taken at Babbitt in the early 1920s.*

DANIEL C. JACKLING *was instrumental in forming the Mesabi Syndicate to explore the commercial possibilities of the taconite lands owned by Williams, Mitchell, and St. Clair. Courtesy American Institute of Mining and Metallurgical Engineers.*

PETER MITCHELL'S MOUNTAIN *was later called Cliff Quarry. From it the Mesabi Syndicate mined and processed taconite during World War I. The first taconite concentrate ever shipped to Eastern steel mills came from this spot.*

A GASOLINE SPEEDER *was the only means of transportation between the Mesabi Syndicate's Sulphur Camp near Cliff Quarry and the outside world in 1916. In front from left to right are E. W. Davis, Anthony F. Benson, Fred Jordan, and Dwight E. Woodbridge.*

THE BUNKHOUSE *(right) and the engineers' shack (center) at Sulphur Camp are shown as they looked in 1919. The camp was located on the eastern Mesabi in a burned-over area near present-day Babbitt, Minnesota.*

IN THE SULPHUR MINE, *operated by the Mesabi Syndicate to supply its Duluth test plant during World War I, taconite was mined with simple tools.*

THE MESABI SYNDICATE'S *test plant at Duluth was used for experimental taconite processing from 1916 to 1918.*

AT SULPHUR SIDING, *taconite rock was loaded into railroad cars for shipment from the Sulphur Mine to the Duluth experimental plant.*

THE SITE *of the Mesabi Iron Company's plant at Babbitt was desolate and remote in 1919. A fire had passed through the rocky country two years before.*

THE MESABI IRON COMPANY *built one-story houses for its employees in the new town of Babbitt. This photograph, taken in 1920, shows the company office under construction. It consisted of five houses like those built for the company's employees joined end to end by short vestibules.*

THE DRAFTING ROOM *at Sulphur was heated by pot-bellied stoves. Here engineers made detailed drawings for the Mesabi Iron Company's mine and plant. The town of Babbitt and the plant were built by the company in 1920–22.*

THE ARGO MINE *was planned to produce 1,200 tons of taconite rock a day for the Babbitt plant. The churn drilling rigs in the foreground made blastholes in the hard taconite at the rate of only about one foot an hour.*

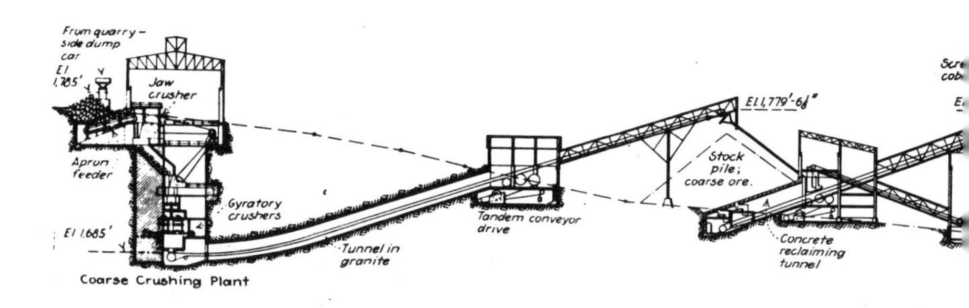

THE MESABI IRON COMPANY'S PLANT *at Babbitt, Minnesota, was the world's first attempt at commercial taconite processing. It opened in 1922 and closed in 1924. During that period, it shipped nearly 150,000 tons of sintered taconite concentrate, much of which was purchased by the Ford Motor Company.*

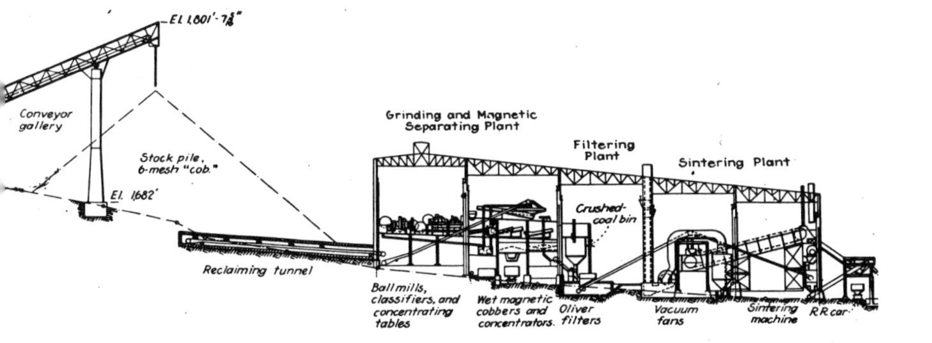
El. 1,601'-7⅛"
Conveyor gallery
Stock pile, 6-mesh "cob."
El. 1,682'
Reclaiming tunnel
Grinding and Magnetic Separating Plant
Filtering Plant
Sintering Plant
Crushed-coal bin
Ball mills, classifiers, and concentrating tables
Wet magnetic cobbers and concentrators.
Oliver filters
Vacuum fans
Sintering machine
R.R. car

THE FIRST *hand-operated Magnetic Tube Separator was developed by E. W. Davis in 1915 at the University of Minnesota's Mines Experiment Station to test samples of taconite sent in by Regent John Williams.*

THE ORE TESTING WORKS *on the University of Minnesota's Minneapolis campus was the home of the Mines Experiment Station from 1914 to 1923 when work on magnetic taconite got under way.*

THE LOG WASHER *(at left below) and a conical ball mill (right) set up in the old Ore Testing Works building in 1916 to test taconite samples from the eastern Mesabi Range.*

THE MINES EXPERIMENT STATION'S STAFF *and visitors about 1916. E. W. Davis is at the extreme left; Dean William R. Appleby is third from the right in the back row; Theodore B. Counselman of the Mesabi Iron Company is on the extreme right; and Henry H. Wade kneels at the right in the front row.*

DEAN WILLIAM R. APPLEBY *was the head of the University of Minnesota School of Mines and the Mines Experiment Station from 1913, when work on taconite began, until his retirement in 1935.*

E. W. Davis *(left), W. G. Swart, an unidentified man, and Ted Counselman were photographed in the new Mines Experiment Station's building about 1924.*

An early wet magnetic cobber *at the Mines Experiment Station. The machine was devised by the staff at the Mesabi Syndicate's Duluth test plant to concentrate crushed taconite particles. It was the most important single development in taconite processing to come out of the Duluth test plant.*

duce Michigan's high-grade ores were quite different from those that produced Minnesota's, and the men who spent their time looking for similar deposits on the Mesabi were disappointed. Mitchell and Chester, among others, did look and were disappointed. Longyear, too, at first sought a formation like that in Michigan, but he quickly changed his mind and, turning westward with the Merritt brothers and others, found the rich hematite and limonite, not in any band or horizon, but as an entirely unexpected type of soft ore occurring here and there at random in the taconite—like raisins in a cake.

Besides clarifying the geological history of the east Mesabi district, the work of Grout and Broderick provided the syndicate with information from which tonnage estimates could be made, and the quality and quantity of the taconite in various areas could be evaluated. The geological study showed that while the taconite between Birch Lake and Mesaba averaged quite uniformly about 30 per cent iron, the proportion of magnetic iron and the size of the magnetite grains varied greatly from place to place and from layer to layer.

Since only the magnetite was found in a sufficiently pure state to be recovered as high-grade ore, all samples secured from natural outcrops or taken from diamond drill cores were sent to the Mesabi Syndicate's newly established laboratory on Thirty-ninth Avenue West in Duluth. There we crushed the samples, ground them fine, and washed them in magnetic tube machines which had been made for the syndicate at the University of Minnesota. From the assay of the tube concentrate and tailings, and from their proportional weights, it was possible to determine the quality and quantity of magnetic concentrate that could be secured from each sample. (See Appendix 1.) The whole property from the Dunka River to Mesaba was sampled, although very little diamond drilling was done except in the Sulphur Creek area in Section 17.

The taconite that the Mesabi Syndicate had leased from the Williams-St. Clair interests consisted of a band about ten miles long from east to west and two miles wide from north to south. At its northern edge, it feathered out to zero thickness where it contacted the underlying quartzite and granite; at its southern edge, at a thickness of several hundred feet, it disappeared under or was swallowed up by a volcanic formation known as the Duluth Gabbro.

There are many things about Mesabi taconite upon which geologists still do not agree. For one thing, great lava flows occurred after the Animikie Sea receded from the Mesabi area and the sediments along its shores and on its bottom had solidified into hard rock. The Duluth Gabbro is one such large volcanic intrusion. It is shaped near the surface like a half-moon, with one horn at Duluth and the other in eastern Cook County. (See Figure 15.) At its northernmost curve, it touches the Mesabi rocks near Mitchell's old Sulphur Creek pit, and it actually cuts off or covers the iron formation east of Birch Lake. Then, farther east, it trends southward, exposing the taconite again in the Gunflint Lake area. The proximity of this great mass of hot rock produced alterations in the original sediments, but the exact nature of the original iron in these sediments, how it got there, and the extent to which the iron minerals were altered, are still being studied. We know that one thing the hot gabbro did was to consolidate the rock, making eastern Mesabi magnetic taconite one of the hardest formations known.

Geologists divide the iron-bearing rocks of the eastern Mesabi into several major layers, the bedding planes of which are approximately parallel to the underlying quartzite and granite. The whole iron-bearing band is called the Biwabik Formation, and its four main subdivisions from the top down are named the Upper Slaty, Upper Cherty, Lower Slaty, and Lower Cherty iron-bearing beds. Below these is the quartzite and on top is the Virginia Slate. The whole formation dips southward at an angle of five to ten degrees under the Duluth Gabbro. Near the edge of the gabbro, the slate is from fifty to several hundred feet thick, and below this the four iron-bearing members are from four hundred to six hundred feet thick. (See Figure 2.)

This information was useful to the syndicate, for in estimating available tonnages, the difficulty of mining and stripping had to be taken into account. Our tests in the Duluth laboratory confirmed those made earlier at the university, showing that on the average about three tons of taconite, after crushing and grinding, would produce about one ton of good magnetic concentrate. Estimates made by the geologists and the syndicate's engineers indicated that it would be possible to mine about 1,500,000,000 tons of taconite between the Dunka River and Mesaba without rock stripping. Some 500,000,000 tons of magnetic concentrate could be made from this amount of rock, but all the concentrate would not be of the same grade. Samples collected from different layers were tested in the tube separator, and it became evident that at the same size of grinding some horizons did not produce as much or as good concentrate as others. We gradually discovered that there was a band near the top of the Upper Cherty horizon which would be difficult to concentrate. This was called "septaria," from a small identifying band running through it.[19]

The geological work at Sulphur satisfied members of the syndicate that there was enough iron in the Williams-St. Clair property to justify large capital expenditures. It now remained for the Duluth experimental plant to demonstrate that high-grade ore could be made commercially from this taconite. The plant, which began operating on June 14, 1916, had a nominal capacity of about a hundred tons of taconite per day. Since it did not run on a continuous basis, however, it probably never processed that much in any one day. From 1916 to 1918 it was operated on a test basis, so that the engineers and metallurgists could learn as much as possible about magnetic taconite.[20]

The plant contained all the various machines that were thought necessary and that were then available for magnetic ore concentration. These machines were operated both separately and in groups, so that we could secure the information needed for designing a large commercial plant and estimating costs of construction and operation. We were instructed to develop processing steps that would produce concentrate assaying 60 per cent iron. As it worked out, the operation of the experimental, or pilot, plant was largely under my direction. Fred Jordan took charge at the mine, and Ted Counselman assisted Swart in the office.

In the Duluth mill, we learned a lot of things about the concentration of magnetic taconite. To crush and grind the taconite rock shipped from the mine, we installed machinery that had been found satisfactory at the western low-grade copper properties with which Swart and Jackling were familiar. But it soon became evident that standard equipment was no match for the hard, abrasive, magnetic taconite of the east Mesabi. We learned that everything it touched would have to be protected against excessive wear with special hard, tough, alloy steels.

Figure 3 shows the "flow sheet," or processing steps, in the Duluth plant as it was originally equipped and operated. Many changes and additions were later made in the equipment and in the methods of operation.

We made careful studies of the possibility of rejecting low-grade nonmagnetic or weakly magnetic pieces of rock at various stages of crushing, and we made elaborate curves and calculations to show the value of this processing step, beginning with rock rejection at sizes as coarse as three inches. We designed magnetic separators, called cobbers and graders, especially for this work. At the finer sizes of crushing, these machines were inefficient and very dusty and disagreeable to operate.

To solve this problem, we developed a wet magnetic cobber that was more satisfactory for ore crushed to a quarter inch in size and finer. In this machine the magnets were encased in a watertight rotating drum; the drum was submerged to about one-third of its diameter in a tank of water. The wet ore was carried by the rotating drum into and out of the water, and the separation of the nonmagnetic from the magnetic ore particles occurred below the water level. The wet cobber has turned out to be the most important single development in taconite processing made in the Duluth plant.[21]

In the fine-grinding operation we demonstrated that while taconite was very hard, it was also, like glass, quite brittle, and under proper conditions could be ground readily. We discovered the value of using small grinding balls but could not apply this principle satisfactorily, because at that time there was no equipment available to crush the taconite fine enough so that small balls could efficiently finish the grinding.

We recognized the fact that after passing over the wet cobber,

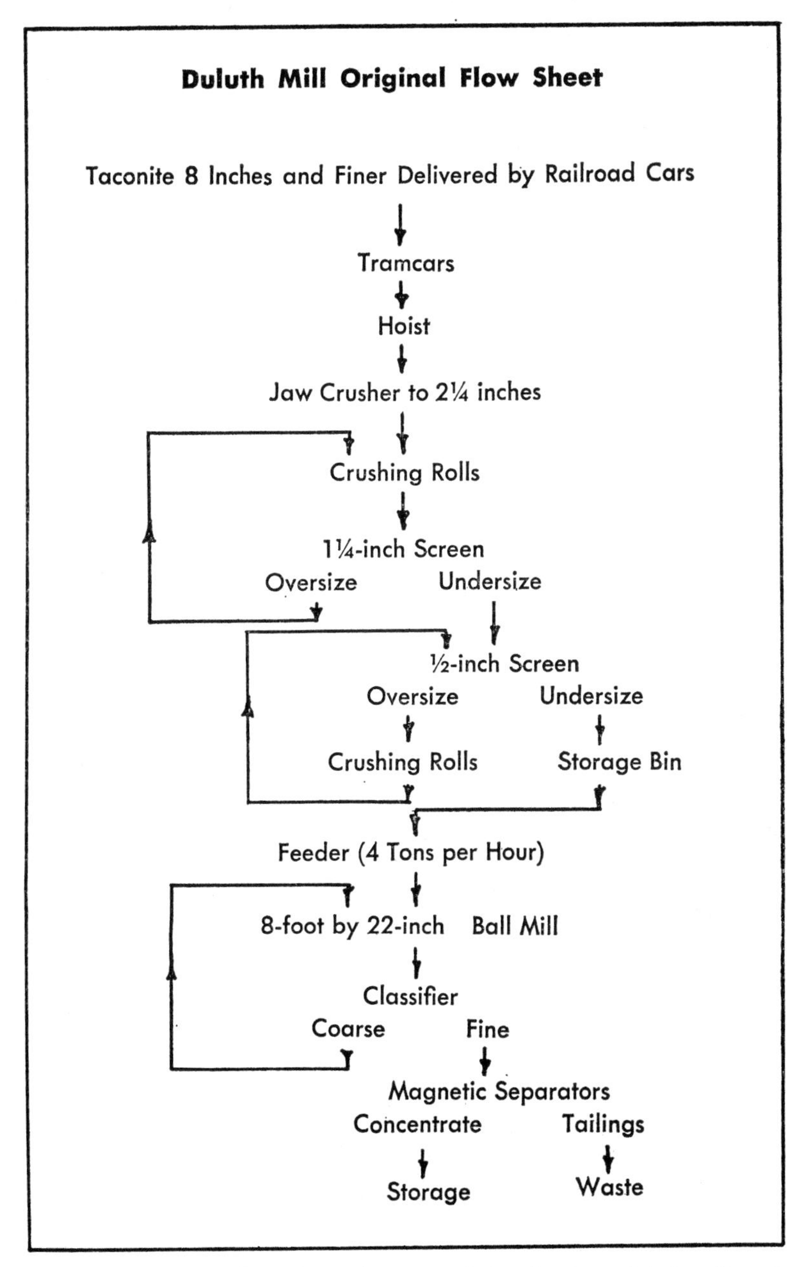

FIGURE 3. *The original flow sheet used in the Mesabi Syndicate's Duluth experimental plant in 1916.*

the particles of magnetite became small permanent magnets and tended to cling together in large masses like snowflakes. This made trouble in the classification steps that followed. We used electric coils that had been developed at the Mines Experiment Station to demagnetize the ore particles and break up the large flakes. However, we did not appreciate the importance of thorough demagnetization at this point in the flow sheet, and so did not put the coils to proper use. As a result, our classification was very inefficient and uncertain.

In the final concentration step, we tried several machines, including the magnetic log washers developed at the university. None of them worked very well at the extremely fine size of grinding required.[22] Removing the water from the fine concentrate was also difficult, but we found that continuous rotary vacuum filters would do the job when the cloth filter covers were maintained properly.

The only agglomerating method we considered was sintering. To the iron and steel industry, agglomeration means gathering a large number of smaller particles into a larger mass sufficiently strong to withstand rough handling. It is ordinarily necessary to agglomerate fine iron ore before it is charged into a blast furnace to prevent it from being blown out the top of the furnace by the heavy blast of air. Any ore particles smaller than 10 mesh are potential blast furnace flue dust, and, therefore, for best furnace practice, such fine particles should be agglomerated in some way. The three agglomerating processes in general use are sintering, briquetting (of which pelletizing is a modification), and nodulizing.[23]

Of these, sintering is the most widely used. In this process, pulverized damp ore is mixed with fine coal. The mixture is then spread evenly on a stationary or a traveling grate, and the surface is ignited. It burns downdraft; that is, the burning zone moves down through the mixture toward the grate and heats the ore in a narrow band to a high temperature. This causes the ore particles to melt and fuse together into strong porous masses resembling clinkers.

In the Duluth plant we did only enough experimental work in this field to determine the capacity of a standard sintering machine and the conditions necessary to operate it. Management had decided that the sintered product was to assay 60 per cent iron. To accomplish this, we found that the final magnetic concentrate should assay about 61 per cent iron, and all the experimental work was aimed at the production of a 60 to 61 per cent iron concentrate.

During the wartime year of 1918, however, we were asked to make as much low-phosphorus concentrate as possible, sinter it, and ship it to the Midvale Steel and Ordnance Company at Coatesville, Pennsylvania, where it would be used in making steel for the noses of armor-piercing projectiles. We had found the rock from one location — Peter Mitchell's mountain, later called Cliff Quarry — would produce very pure concentrate high in iron and low in phosphorus, an impurity that must not be allowed to contaminate high-quality steel. Ores being shipped down the lakes in 1918 contained on the average 0.104 per cent phosphorus, but from the Cliff Quarry outcrop, we could make concentrate having only 0.008 per cent phosphorus and 63.23 per cent iron.[24]

We received orders to put the Duluth plant into continuous operation and produce as much low-phosphorus sinter as possible. To do so, a crude sintering plant was built at Thirty-ninth Avenue. It was a homemade affair, and its operation was haphazard. Nevertheless, it has some claim to fame, for before the end of World War I it produced 1,840 tons of sintered concentrate which were shipped down the Great Lakes to the Midvale Steel and Ordnance Company — without doubt the first taconite concentrate ever shipped to Eastern steel plants.[25]

The operation of the Duluth plant attracted considerable attention in the iron and steel industry, and many visitors came to see it and to discuss taconite utilization.[26] With D. C. Jackling as its champion, low-grade magnetic ore began to take on new importance. Other magnetic properties were brought to the Mesabi Syndicate's attention, and many samples were tested in the Duluth laboratory. Eventually, Swart, acting for the syndicate, took leases or options on several other properties.[27]

Throughout the period of experimental operation at the Duluth plant, Jordan, Counselman, and I prepared reports on our progress in developing a flow sheet and the equipment to be used eventually, we hoped, in a commercial plant.[28] Later many detailed estimates of the cost of constructing and operating plants of different sizes were worked out using various flow sheets. From these reports as well as data secured by personally inspecting other plants, Swart provided Jackling and the Mesabi Syndicate with information that would enable its members to decide whether or not commercial taconite processing was warranted at that time.

By 1919 the syndicate's members had spent about $750,000 on taconite experimental and developmental work at Duluth and Sulphur.[29] They had obtained geological information which demonstrated that sufficient taconite was available on their lands to supply a commercial operation. Their test plant at Duluth had established the fundamentals of taconite processing and had developed and standardized a method of determining the analysis of magnetic iron that is still in general use. The syndicate had also shipped the first taconite concentrate ever received by Eastern steel mills.

In December, 1918, when the experimental work was finished, I left the Mesabi Syndicate to return to the Mines Experiment Station. I still acted as a consultant for the syndicate, however, and spent much time at Duluth and Sulphur. All during the summer and fall of 1919 the syndicate's members considered the results of the project. They could choose one of two alternatives: to abandon the entire effort and give up the lease on the eastern Mesabi lands, or to move ahead toward commercial production. Near the end of the year, they decided to organize an operating company and provide the necessary additional funds for a commercial plant.

3 The First Commercial Taconite Plant

LATE IN 1919 Swart was instructed to proceed with the construction of a commercial taconite concentrating plant near Sulphur. Only one unit was to be built, but the site was to be so chosen and the plant so designed that operations could be expanded if this became desirable. The initial unit was to have a capacity of about 400 tons of shipping product per day or 120,000 tons per year. The syndicate did not anticipate that this small plant would be highly profitable, but it was expected to demonstrate the metallurgical and financial feasibility of commercial taconite processing. If successful, it would act as the forerunner of additional units that would make the project financially attractive. It was estimated that the entire project, including a town, a power plant, and all auxiliaries, would cost approximately $3,000,000, and that the finished taconite concentrate in the form of a sintered product would be worth about $7.00 a ton at lower Great Lakes ports.[1]

In November, 1919, a new firm called the Mesabi Iron Company was organized. Charles Hayden took over the somewhat involved financing of the venture. Members of the Mesabi Syndicate were given the right to subscribe for preferred stock with a common stock bonus. The officers of the company were: Hayden, chairman of the board; Jackling, president; Swart, vice-president and general manager; William G. Sargent, vice-president; John R. Dillon, treasurer; and Arthur J. Ronaghan, secretary. The last three men were members of the Hayden, Stone firm.[2]

With the formation of the Mesabi Iron Company in 1919, the Mesabi Syndicate passed out of existence. The Duluth plant was dismantled, and the machinery was either sent to Sulphur or sold. A large bin filled with more than a thousand tons of concentrate occupied a space across Thirty-ninth Avenue just east of the plant. In 1920 this was sold to the Jones and Laughlin Steel Corporation.[3]

After the decision had been made to build a commercial taconite plant on the east Mesabi, Swart's first job was to assemble a group of men to design and engineer the construction. He decided to do the engineering at Sulphur and had a new office and drafting room built there near the camp. Some of the old buildings were used as storehouses, bunkhouses, and a mess hall.

The work of selecting the best sites for the new mine and plant was greatly simplified by a fire that had passed through the area in 1917. It had burned all the "down timber" and in many places even the moss that covered the rocks, making it possible to walk over the country at will. The old trails and survey lines were no longer the only avenues of travel.

A spot just south of Sulphur Camp near the old Peter Mitchell pit seemed the logical place to locate the mine, for it had a large exposure of taconite that had been drilled and sampled and was known to be of a quality which could be comparatively easily concentrated. This site was covered in only a few places with overburden which would have to be stripped off to reach the ore. More information was needed, however, about the exact nature of the taconite, and for this reason diamond drills were brought into the area to sink shallow holes at close intervals.[4]

While this work was going on, the engineers were studying possible places to build the plant. A number of requirements had to be met, perhaps the most important of which had to do with the water supply and the disposal of tailings from the concentration process. For one thing, the Sulphur area was almost exactly on the divide between the Hudson Bay and Lake Superior drainage basins, and laws prohibited taking water from one basin and returning it to another. Jackling came up to

Sulphur in his private railroad car, which was named the "Cypress," and after looking over some of the proposed locations, he decided to erect the mill on the spot where the present Babbitt plant of the Reserve Mining Company stands. It was situated on the north slope of the Giant's Range about a mile and a half north of the newly selected mine location. The hills there slope down sharply to a swampy area, which drains into Birch Lake about two miles away. Water for the plant was to be pumped from Birch Lake; the wet tailings were to be deposited in the swampy area between the plant and the lake.[5]

As soon as the millsite was chosen, the engineers began to lay out the mine and the railroad. It was decided to mine the taconite by methods then used in some New York traprock quarries. A sloping excavation would be started as a cut, giving a face about twenty feet high. Then the cut would be widened to form a mine pit, with a working face about a hundred feet long. Some fifteen feet behind the edge of the face a row of holes would be drilled about fifteen feet apart. They would be loaded with dynamite and the taconite in front of them blasted down into the pit. The holes would have to be so spaced and loaded that the taconite would be broken fine enough for the steam shovel to handle; on the other hand, the dynamite charge could not be so large that it would blow the rock all over the pit.

But how were the blastholes to be drilled into the hard taconite? Various types of drills had been tried with only limited success. Swart and Fred Jordan finally decided that churn or well drills were the best available tools. After this decision was made, the exact design of the drill bit came in for much study and discussion. The choice involved not only the size and shape of the bit and its cutting edges, but also the type of steel from which it should be made. The bit would have to resist wear and breakage, but it must be easily sharpened and tempered by a blacksmith using a forge and hand tools.

The bits that seemed best were about four feet long and five inches in diameter, with a single cutting edge like a blunt chisel. In use, they were attached to a drill stem — a piece of round steel some four inches in diameter and twenty feet long — and the whole assembly weighed approximately a thousand pounds. The

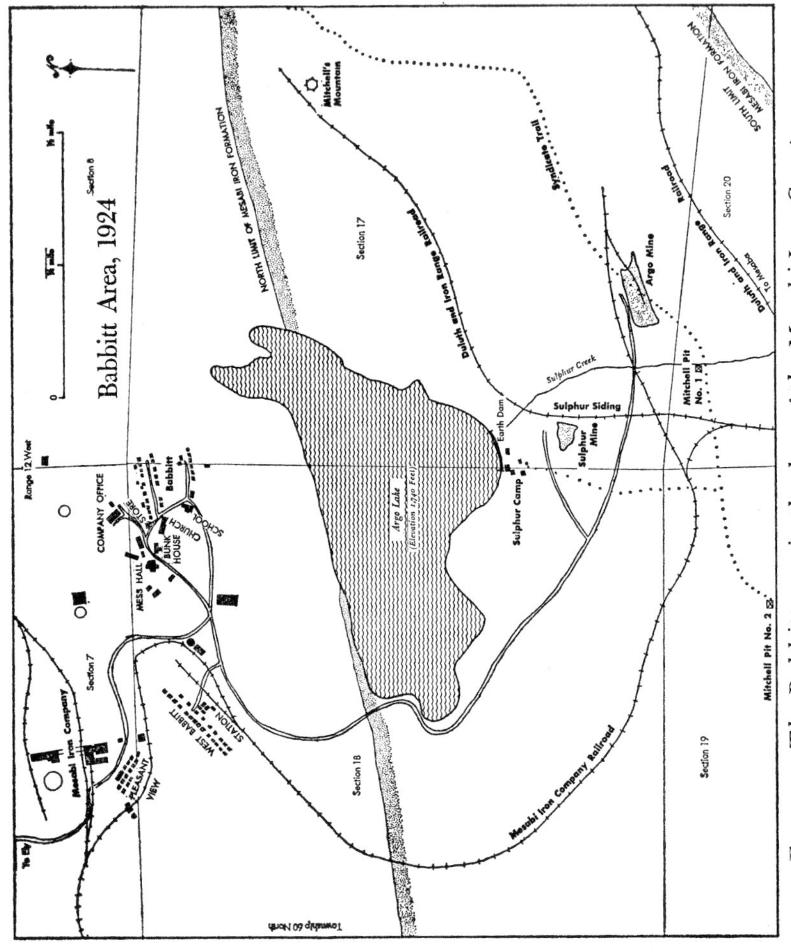

FIGURE 4. *The Babbitt area in the days of the Mesabi Iron Company.*

drilling machine raised this assembly about four feet and dropped it some forty times a minute, the full impact of the blow driving the chisel edge of the bit against the taconite at the bottom of the hole. The drill stem and bit were turned slightly between blows, and as the taconite was gradually chipped and worn away, the hole slowly became deeper. Occasionally a bucket or two of water was poured into the hole. At regular intervals the drill would be withdrawn and a bailing bucket would be lowered into the hole to remove the taconite chips and mud.[6]

Experimental drilling of this hard taconite rock was exceedingly slow, the average speed being about a foot an hour. At the end of that time, it was necessary to remove the drill bit and take it to the blacksmith to be sharpened. This was before the day when temperature-controlled heating and tempering furnaces or power-sharpening tools were in general use, and whether a bit was properly sharpened and tempered depended entirely upon the skill of the blacksmith. He was equipped with a coal-fired forge, an anvil, and a barrel half full of water for tempering. He had some mysterious white powder that he kept in little papers in his pocket, and once in a while he would put some of it into the barrel of water. This was supposed to harden the steel. Something did, but no two bits were alike. On the average, the bits could be expected to drill twelve inches into the taconite before sharpening and tempering were necessary, but one bit might drill twenty inches and the next only four inches before it became dull.

All these facts had to be taken into account by the engineers planning the operation of the mine and determining the equipment required. To supply a plant with a planned capacity of 400 tons of concentrate a day, the mine would have to produce 1,200 tons of rock daily. The flow sheet developed in the Duluth experimental plant was to be followed in designing the new operation,[7] and as a guide for the engineers, the following steps and conditions were established:

1. The mine-run taconite was to be crushed in three stages to pieces two inches in size.

2. The two-inch taconite was to be crushed to one-fourth

inch in a closed circuit with rolls, using dry magnetic cobbers to reject the low-grade particles at various sizes between two inches and one-fourth inch.

3. Wet cobbers were to be used to reject low-grade particles finer than one-fourth inch.

4. The one-fourth-inch wet cobber concentrate was to be ground in two stages to 150 mesh.

5. It was then to be concentrated magnetically to a 60 to 61 per cent iron content. The weight recovery of the concentrate was expected to be about one-third the weight of the original taconite.

6. The fine concentrate was then to be dewatered in continuous filters, mixed with about 7 per cent of its weight of fine anthracite coal, and sintered.

7. The sinter was to be loaded into railroad cars for shipment to Two Harbors on the Duluth and Iron Range Railroad and from there sent by boat to lower lake ports. In winter, the sinter was to be stock-piled near the plant for shipment the following summer.

8. The sinter was to be sold as Old Range Bessemer grade assaying 60 per cent iron natural and 0.030 per cent phosphorus. (See Appendix 1.)

9. Water for processing the ore was to be pumped from Argo, Iron, and Birch lakes into a reservoir near the plant. The total water requirement was to be about thirty to thirty-five tons for each ton of concentrate.[8]

10. The dry cobber tailings were to be hauled away and sold as crushed stone for concrete aggregate, road building, or railroad ballast. The fine wet tailings were to flow by gravity from the plant into the large swampy area between the plant and Birch Lake.

With this information and complete surveys of the area, the engineers were in a position to make detailed drawings of the whole processing plant and railroad. Actual construction got under way early in 1920. The first step was to extend the Duluth and Iron Range Railroad spur to the millsite in order to bring in construction materials and workers. An automobile road of sorts ran from Aurora through Embarrass to Ely, but there was no connection to the millsite or the Sulphur Camp area; the only access was still by railroad sixteen miles to Mesaba. The plant, townsite, and railroad were built without trucks, bulldozers, or other heavy equipment. Horses with wagons and scrapers, and men with shovels and wheelbarrows, did the grading and made the excavations. It was slow, hard work, and the construction gangs were made up of tough men.

At first the crews worked a week, took the train to Duluth on Saturday afternoon, spent their money, and returned broke on Monday. Then with the coming of Prohibition in 1919 the whole pattern changed, because the men could not spend their week's wages between Saturday and Monday. They tried, but a man can drink only so much milk, and without the help of strong drink the ladies were not so alluring. As a result, many of the men did not make it back to camp the following Monday, and the work was seriously slowed down. Prohibition may have been a fine thing for the nation, but it did not help Swart and Jordan get the Mesabi Iron Company's plant built. They tried to keep the men in camp by scheduling lectures for them, holding church services on Sundays and Thursdays, and bringing in a circulating library. The latter effort was a failure because most of the workmen could not read English, or even understand it very well, and did not want to learn.

When Jackling came to inspect the job, he was much disturbed by the lack of progress. In no uncertain terms he expressed his impatience with Swart and Jordan for allowing the men to get away from camp. We all knew the stories of how Jackling built his great plant in Utah by using some buildings in a nearby dry gulch to provide wine, women, song, and roulette tables. "Feed 'em and keep 'em broke," was his motto, and it worked.

Not long after one of Jackling's visits two enterprising Finns from Embarrass, so the story goes, moved into some abandoned lumber camp buildings near the mouth of the Dunka River and brought with them some "girls," strong drink, and a card game. From then on Swart had no trouble keeping the men on the job. It was three miles from the camp to the mouth of the Dunka, but there was soon a beaten trail all the way. One of

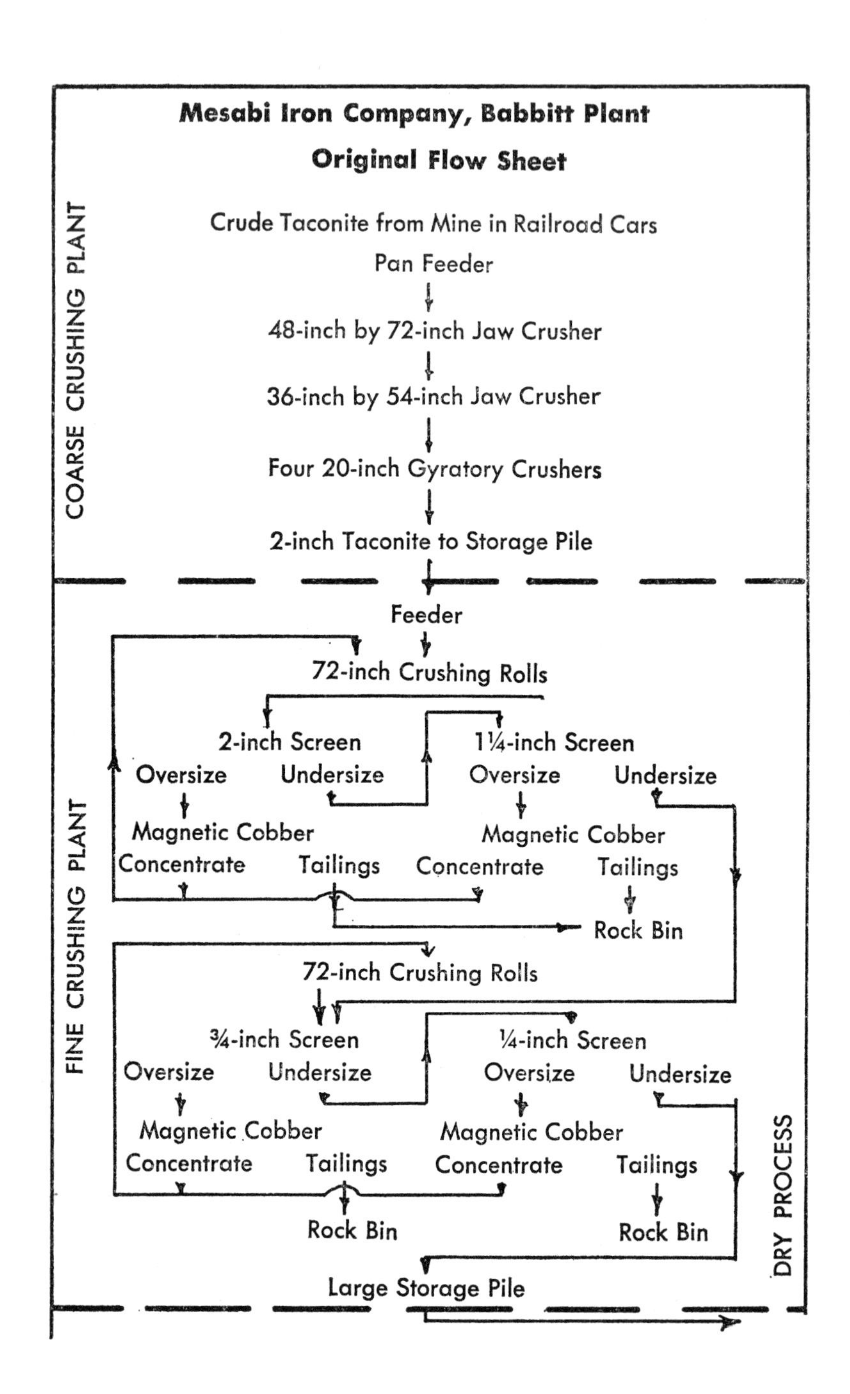
Mesabi Iron Company, Babbitt Plant
Original Flow Sheet
COARSE CRUSHING PLANT
Crude Taconite from Mine in Railroad Cars
Pan Feeder
48-inch by 72-inch Jaw Crusher
36-inch by 54-inch Jaw Crusher
Four 20-inch Gyratory Crushers
2-inch Taconite to Storage Pile
FINE CRUSHING PLANT
Feeder
72-inch Crushing Rolls
2-inch Screen
Oversize
Undersize
1¼-inch Screen
Oversize
Undersize
Magnetic Cobber
Concentrate
Tailings
Magnetic Cobber
Concentrate
Tailings
Rock Bin
72-inch Crushing Rolls
¾-inch Screen
Oversize
Undersize
¼-inch Screen
Oversize
Undersize
Magnetic Cobber
Concentrate
Tailings
Magnetic Cobber
Concentrate
Tailings
Rock Bin
Rock Bin
Large Storage Pile
DRY PROCESS

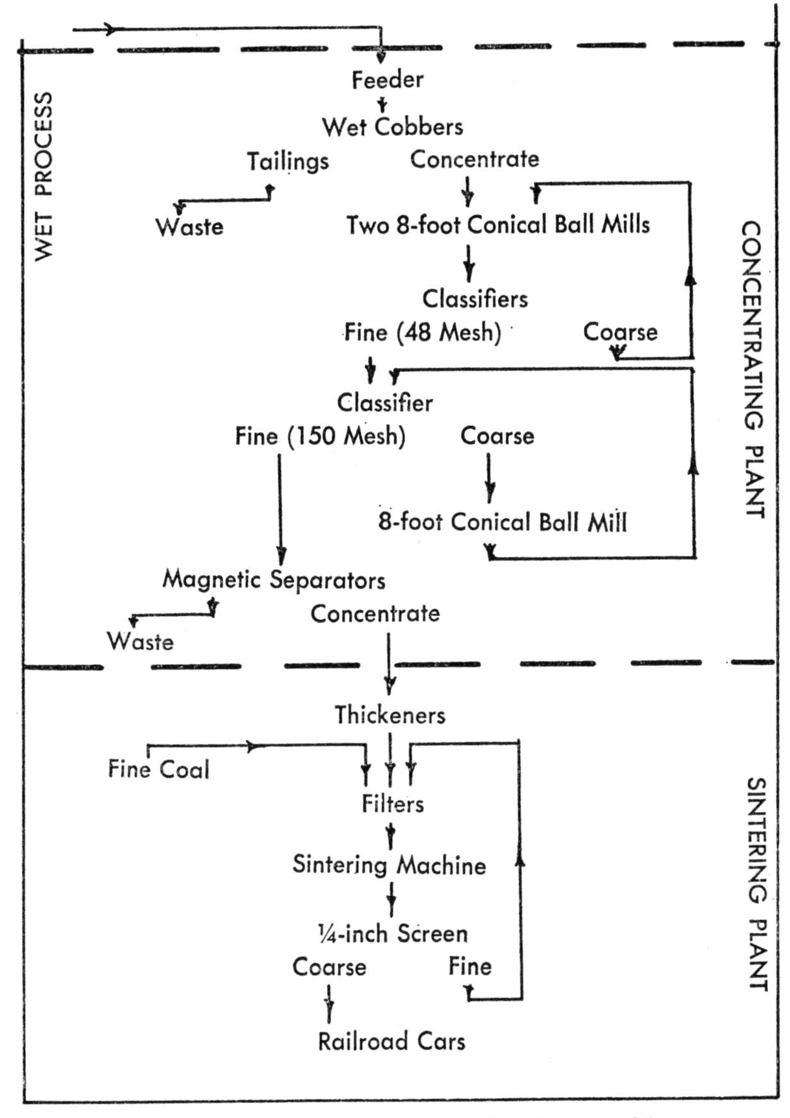

FIGURE 5. *The original flow sheet used in the Mesabi Iron Company's Babbitt plant in the early 1920's to produce a sintered product. Many changes and modifications were made in the arrangement of this equipment before the plant closed in 1924.*

the women, a buxom blonde, was called Sandpaper Annie. For her companionship she collected one dollar in silver; no paper money for her. She kept her money in a nail keg with a slot in the top. It was nailed to the floor but was pretty safe anyway, because in a short time it was so heavy no one could have carried it very far. Silver dollars got to be as scarce as hens' teeth on the eastern Mesabi and were selling at a premium as far west as Mountain Iron.

Everything went fine for a while. The men were happy, the plant was going up, and even Jackling was pleased with the progress. But eventually some do-gooder reported to the authorities that a "blind pig" was operating at the mouth of the Dunka (he didn't know the half of it), and the revenue agents came, ran the people out, and burned the place down. By that time, however, the plant was nearly finished.[9]

While work was going ahead on the plant and the mine from 1920 to 1922, a town for the company's employees and their families was also being built under Swart's direction. It was situated near the plant and about a mile and a half from the mine. Many locations were considered, including the site of present-day Babbitt, but automobiles were rare in those days and it was necessary for the employees to live within walking distance of their work. A spur track was built to the townsite and a railroad station established. It was used mostly for freight, but a passenger car was attached to the freight train, which at first came in two or three times a week and later every day.

The houses built by the company in the new town were all much alike — one story and without basements. Each had two bedrooms, a kitchen, a living room, and bath. When a larger building was needed, two of these standard houses were simply combined, end to end, and the partitions were changed. For the company office, several of the structures were put together with short vestibules. A dormitory was erected to house the single men and visitors. It had two stories and about twenty rooms, including a sort of lobby or living room at one end for poker games. There was only one other two-story building in town — the general store. It had living quarters for the storekeeper and his family on the second floor. The town was provided with a general community building—also used as a church—and it had a small hospital presided over by Dr. Paul J. McCarty, who practiced in Mesaba and later in Ely. He took care of everything from children's diseases to plant accidents and gunshot wounds. Serious cases were sent to the hospital at Two Harbors.

For the construction crews there was a bunkhouse, forty by one hundred feet in size, and near it a mess hall in the form of a star. From the kitchen at the junction of the wings, food was prepared and served to about four hundred men. Everyone ate at the same time. When the cook rang the bell for meals, the food was already on the table, and if we did not hurry over immediately everything was cold. If we did rush over, we were in danger of being trampled. It was quite a thrill to eat there occasionally, but as a steady diet the mountains of food and the smell of frying mingled with that of unwashed men soon dispelled the glamour.

Providing underground water and sewer pipes for the town would have been an expensive undertaking in the severe climate of northern Minnesota. Instead of laying them in the solid rock the company put the pipes on the surface, boxed them in, insulated them, and heated them by circulating the village water through a boiler during freezing weather. For the first few months after the town was built, its water came from Argo Lake, which had been artificially created by damming the outlet of the swamp between Sulphur Camp and the new town. Great numbers of minnows had taken over this new lake, with the result that dead minnows were forever getting into the water pipes and appearing in strange, unappetizing places such as bathtubs or dishpans. Drinking water from a small well was delivered by a tank wagon until a municipal well was drilled, a 50,000-gallon water tank erected, and a filtration plant installed. These were not completed until the project was well advanced. We did have bathrooms and electric lights, however, and no one can properly appreciate these conveniences who has not lived in a cold climate without them.

The new town was at first called Argo, for the ship used by Jason and the Argonauts in their search for the golden fleece, but later it was changed because there was another town in the state by that name. It was rechristened Babbitt—not as some people think for the character in Sinclair Lewis' *Babbitt*—but in honor of Judge Kurnal R. Babbitt of New York City, a long-time legal adviser of Hayden and Jackling who had just died.[10] The name Argo, however, stuck to the township, the cemetery, and the little lake between Babbitt and the mine.

I think the first employee to move to Babbitt with his family was August Palo, foreman of the experimental plant at Duluth, who took up residence there with his wife and two children in 1920. The town grew as more men were added to the staff; more houses were built, and two new additions—called West Babbitt and Pleasant View—came into being. The company built no houses in West Babbitt. Instead it sold building materials at cost and leased the land to those who wished to erect their own homes. Most of the houses were made of rough lumber and tar paper, and West Babbitt was not exactly a show place. Pleasant View, west of the mill at the top of the hill overlooking the Embarrass Valley, was a company-built residential area of standard houses, primarily for single men. From this spot the lights of Tower could be seen, and on clear nights those of Ely were also visible. A company mess hall was also erected for the men living in this community. The Babbitt townsite was always regarded by the company as temporary, and the houses were built with the idea that they would be moved to a permanent location at some later date. By 1924 there were twenty-five dwellings in Babbitt, twenty-eight in West Babbitt, and sixteen in Pleasant View.[11]

When the town of Argo was organized in 1920, its population totaled ninety-eight persons. A year later it had an estimated five hundred inhabitants, and the village of Babbitt had about four hundred. A school district (number 83) was organized in 1921, and the company built a two-room elementary school at Babbitt, which was attended by sixty-six children in 1923. This would have been about one child per every four regular employees, since the company at the peak of its operations early in 1924 employed about 245 men.[12]

When the first election was held in 1920, employees of the

Mesabi Iron Company filled most of the offices. Manager Swart, mine superintendent William Mudge, and comptroller O. C. Burrell were elected town supervisors; Mrs. James Mitten, wife of the company's chief chemist, was clerk; Clyde M. Pearce, superintendent of the sintering plant, was named assessor; Dr. McCarty was treasurer; Ted Counselman and William J. Baumgrass, the firm's accountant, were elected justices; and Oscar Birkness and Al Johnson became constables.[13]

As a new town arose on the bare rocky hills, the mill it was to serve also neared completion. Late in 1921, however, a short time before the plant was ready to go into operation, the company discovered a startling flaw in its plans. It seemed that the 60 to 61 per cent iron sinter for which the plant had been designed was not good enough to interest steel producers. Instead 64 to 65 per cent sinter was required. About that time Clement K. Quinn of Duluth, an experienced Lake Superior iron mine operator and sales representative, was appointed to handle the Mesabi Iron Company's sinter. It was his job to sell the product and arrange for its transportation from the plant to the furnaces of the purchasers. Early in 1922 he published a sales leaflet guaranteeing that the sinter would contain 64 per cent iron natural.[14]

No sinter of this quality had ever been made in the Duluth experimental plant; moreover, the objective of all the development work had been to make a 60 to 61 per cent iron product. In his 1915 report to Jackling, Swart had stated: "From all the results I have been able to get on the separation of these ores, I believe a concentrate assaying 60 to 62% iron can best be made with about .025 phosphorus. It goes without saying that this is a very desirable product for the furnace man of 1915. . . . It would be possible to make a concentrate assaying upward of 65% iron, but the demand for such extremely clean material seems likely to be too limited to warrant the expectation of the necessary premiums, and it is much safer to make all calculations on the basis of 60% to 61% concentrate, the market for which is wider and assured." Swart also suggested that "Before spending very much money, it would seem wise to make sure . . . that the sintered concentrate can actually be sold under contract for prices given us." Unfortunately this was not done until the plant was nearly ready to operate.[15]

It is necessary at this point to consider the quality, value, and price of iron ore, because incorrect conclusions reached on these important matters caused the problem that faced the Mesabi Iron Company. The firm had assumed its 61 per cent iron sinter would be a premium product commanding a high price. In 1921 it was discovered that the pricing formula which the company had used applied to Lake Superior hematite and that it applied only with modifications to the magnetite sinter that would be produced at Babbitt.

The method used to arrive at the price of Lake Superior iron ores is complicated, and it is easy to see how an error of this kind could be made by people inexperienced in the business of buying and selling such ores. Early each year what is known as the "Lake Erie Base Price" is established. It varies from year to year, is determined in the spring by the first sale of any considerable tonnage of ore, is published in the trade journals, and is accepted by the industry as the base price for the following shipping season. From it, by using a prescribed formula, the prices of ores of various qualities can be computed for a given year. Such ores are classified as "Mesabi" or "Old Range" and as "Bessemer" or "Non-Bessemer."[16]

The ores mined on the Marquette and Gogebic ranges of Michigan and on the Vermilion in Minnesota — the so-called Old Range ores — were predominantly hard and lumpy; it has been customary for blast furnace operators to pay about 25 cents more a ton for them than for the fine, earthy ores of the Mesabi. In effect the Old Range classification offers a structure premium. Since the Mesabi Iron Company's sinter would contain very little fine material, it was decided to attempt to sell it as Old Range ore. The Bessemer or Non-Bessemer classification was determined by the phosphorus content of the ore. If it was under 0.045 per cent, the ore was classed as Bessemer, and some years received a premium of from 15 to 75 cents a ton over Non-Bessemer. Since the Mesabi Iron Company's sinter was expected to contain only 0.025 per cent phosphorus, it was thought that it would be classed as Bessemer grade. The final classification under which the firm expected to sell its product, then, was Old Range Bessemer, making the sinter eligible for both premiums. In addition, premiums were allowed at that time on a sliding scale for phosphorus below 0.045 per cent and for iron above 55 per cent, and the company also expected to take advantage of these.[17]

Unfortunately "Mesabi Sinter," as the company's product was called, had a relatively high silica content, and therefore it was not the premium product the company's officials thought it would be. In 1921 the average analysis of ore shipped down the lakes was 52.07 per cent iron natural, 8.23 per cent silica, and 10.95 per cent moisture. The Babbitt plant was designed to produce concentrate assaying 61 per cent iron (dry and natural), 12.85 per cent silica, and no moisture. Although the sinter's higher iron content appeared to make it the more desirable product, it was not as good chemically, from the blast furnace operators' viewpoint, as average lake ore. To produce a ton of iron from average lake ore, the furnace operator would have to flux, melt, and waste as slag about 300 pounds of silica, while with Mesabi Sinter he would have to waste over 400 pounds, thus considerably increasing his smelting costs. For Mesabi Sinter to be merely the chemical equivalent of average lake ore in the blast furnace, the silica content would have to be reduced to 9 per cent and the iron content increased to about 64 per cent.[18]

As the company's sales representative, Mr. Quinn found that unless its product could be improved, the sinter would not only be ineligible for premiums but would be difficult to sell without accepting a severe penalty below the computed price. Management proposed to meet the problem by reducing production, grinding finer, washing harder, and, if necessary, discarding a somewhat higher grade of tailings. By sacrificing tonnage and metallurgy, the company hoped to increase the grade of its sinter from the 61 per cent for which the plant was designed to the 64 per cent that Mr. Quinn had guaranteed in his sales leaflet. On the basis of this guarantee, in June, 1922, Mr. Quinn had sold a large order amounting to 60,000 tons of sinter to the Ford Motor Company to be shipped to its facilities at River Rouge, Michigan.[19]

The first plant in the world to attempt taconite processing on a commercial scale went into operation at Babbitt on June 21, 1922. The rate of production was very low at first, as was to be expected in a new plant. Most of the men had never seen the machines they were supposed to run and had to be taught to operate them. It was frequently necessary to shut down for repairs and alterations, occasionally for a week at a time. But the plant did operate. Taconite rock was fed continuously into one end of it, and sintered concentrate came out at the other. Although everything did not function as well as we had hoped, it worked well enough to prove that, as planned, 400 tons of sinter per day assaying 61 per cent iron could be produced. Whether 64 per cent sinter could be turned out in sufficient quantity remained to be seen.[20]

The Mesabi Iron Company's first cargo was shipped on October 1, 1922. It totaled 5,076 tons, assaying 62.02 per cent iron natural, 0.028 per cent phosphorus, 9.34 per cent silica, and 2.59 per cent moisture. It was shipped in standard ore cars to Two Harbors, loaded into the steamer "Central West," and consigned to the Ford Motor Company. A second cargo of 5,600 tons went to Ford in November. Thus total shipments for the first year amounted to only 10,676 tons.[21]

Early in May, 1923, the largest single shipment of sinter by the Mesabi Iron Company — 22,265 tons assaying 61.20 per cent natural iron — went by rail to the Ford company. Severe complaints were received from Ford about the quality of this sinter. Quinn reported to Swart that Ford "would refuse to accept any material carrying less than 62.00 per cent iron." On July 9, 1923, in a letter to Swart, Quinn wrote that the company had in the Two Harbors ore dock "one block of 6,600 tons carrying 61.22% iron and 12.27% silica." He said that he planned to hold this until the plant could ship some higher-grade material to blend with it to make up a cargo that "will be sufficiently high in its analysis to qualify under the contract." Swart replied on July 12: "I understand the gravity of the situation and . . .

am doing everything I possibly can to get and keep the grade of the sinter up."[22]

While the purity of the product the plant was able to make was above that for which it had been designed, the quality was not high enough to meet the guarantee. The plant's best record, from the analysis standpoint, was made in May, 1924, when a cargo shipped to Ford totaled 5,656 tons and assayed 64.39 per cent natural iron, 0.032 per cent phosphorus, and 8.38 per cent silica. To do this, a new so-called Davis Flow Sheet was used, which included table concentration.[23] (See Figure 6.) In order to increase the iron assay, the tables rejected relatively high-grade tailings and, therefore, reduced the weight of concentrate recovered from each ton of crude taconite. This, of course, increased operating costs, making the whole project less attractive to Jackling and the board of directors.

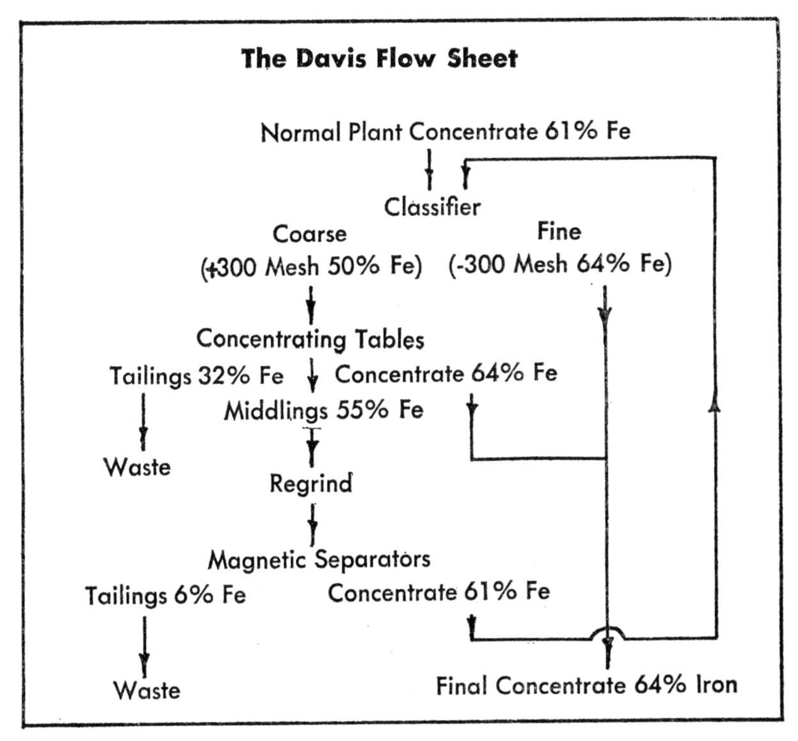

FIGURE 6. *The "Davis Flow Sheet" used in the Babbitt plant to produce a higher-grade product containing 64 per cent iron.*

As long as the plant operated, we continued the struggle to boost capacity to 400 tons a day and hold the assay of the sinter at least close to the guarantee. The best period of operation from a tonnage standpoint came in the spring of 1924, when production rates as high as 432 tons of sinter a day were reached occasionally. By that time modifications in the equipment had made it possible for the plant to produce quite regularly 350 to 400 tons of sinter a day assaying about 62 per cent iron and 10 per cent silica. By reducing the production rate to 250 or 300 tons a day and using the Davis Flow Sheet, the plant could produce sinter assaying slightly over 64 per cent iron and 8 per cent silica. Gradually it became evident that the Mesabi Iron Company's plant could produce the higher-grade product required to satisfy the blast furnace operators or the higher tonnage required to satisfy Jackling and the directors, but it could not do both at the same time.

Swart had estimated in 1919 that 61 per cent iron sinter could be produced in the Babbitt plant at the rate of 364 tons a day for $6.45 per ton. The plant was built on the basis of this estimate. In 1921, because of increases in the cost of labor and supplies, a new estimate indicated that a 364-ton plant would produce sinter for about $7.65 a ton. In April, 1924, it did produce sinter for $7.30 a ton, which was better than anticipated. Although the small plant operated at a loss, this had been expected, and in general the operation lived up to Swart's 1921 predictions.[24]

The general manager had also estimated that a larger plant, which could produce 800 tons a day, would pay its own way or perhaps show a little profit. In 1923 the directors decided to enlarge the plant to this capacity. Plans were drawn, new machinery was ordered, and some of it had been delivered, when in the spring of 1924 the Lake Erie Base Price of Old Range Bessemer ore dropped 80 cents a ton below the 1923 price of $6.45. This meant that the selling price of Mesabi Sinter fell below the point at which even a larger plant could operate profitably. The whole project was, therefore, as Jackling put it, "suspended until such time as either higher prices for the product or lower costs . . . leave a fair margin of profit."[25]

The 1924 price cut convinced the company's directors that the profitable operation of Mesabi Iron depended upon competing ore producers who were in a position to establish the Lake Erie Base Price. With this knowledge, the directors' enthusiasm for the venture waned; they had learned about the iron ore business the hard way. The obvious remedy would have been to negotiate a long-term contract or to make a working agreement with some ore-consuming company. As early as 1923 it was rumored that Henry Ford had bought into Mesabi Iron, but this was denied and nothing ever came of it, which was perhaps too bad.[26]

Mining was suspended about May 1, 1924. The plant continued to operate at reduced capacity until June 4, 1924, cleaning up the various bins of ore on hand. After that, the sinter remaining in the stock pile was shipped, so operations did not cease entirely until June 10, 1924.[27] A few men were retained to oil and paint the equipment to protect it against damage from rust or exposure. The village of Babbitt was rapidly abandoned, and by the fall of 1924 it had become a ghost town. One of Swart's assistants, Frank A. Emanuelson, remained to take care of the office and to supervise the watchmen and caretakers of the property.

Many simple reasons for the failure of the Mesabi Iron Company's taconite project have been advanced, such as the hardness of the rock, the high cost of drilling blastholes, the difficulty of sintering the concentrate, and so forth. The generally accepted explanation has been that the project was "ahead of its time," that there was still too much high-grade, low-cost ore on the Mesabi Range with which taconite could never compete. It would seem, however, that the primary cause of the firm's difficulties was the failure of its management to evaluate properly the trends in the market for iron ore. Not until 1948 did the Lake Erie Base Price of Old Range Bessemer again get above its 1923 price of $6.45 a ton.[28]

The drop in ore prices was caused by decreased demand and oversupply. Some overexpansion had occurred in the iron mining districts; open-pit, direct-shipping properties had been stripped and put into production without adequate available markets. Thus considerable "distress" ore was thrown on the market, ore upon which development costs, advance royalties, and taxes had been paid and which had to be sold to recover investments. Yearly shipments from Minnesota, while varying greatly, had gradually increased from over 9,000,000 tons in 1900 to more than 46,000,000 tons in 1916. Then came a leveling-off period, followed by a great decline to a little over 2,000,000 tons in 1932.[29]

It is possible now to look back and see that during the early 1920s, when the Babbitt plant was being projected and built, great changes were taking place, or were approaching, in the iron ore and steel business. Large consolidations of operating companies, with contracts to supply the ore requirements of particular steel firms, were in progress. The days of the small, independent producer who sold his ore "over the counter," so to speak, were about finished. Steel companies, with their enormous investments, were feeling the need to control their own ore supplies. While this trend gave stability to the ore-producing industry and was a fine conservation measure — since a company operating several mines could utilize the lower-grade ores by mixing them with rich ones — the consolidation process greatly reduced the market for the independent producers. The independents were gradually absorbed by larger organizations operating as agents for particular steel companies. The Mesabi Iron Company, an independent organization owned by a group of stockholders and not affiliated with any particular ore consumer, had to sell its product wherever it could find a buyer. Only a limited number of purchasers of independent ore were left.

It now seems apparent, too, that the product which the Mesabi Iron Company's management decided to make could not be easily marketed. Some of the qualities that the firm had considered important for its sinter in 1915 were found to be of little value in 1923, and other qualities that had received no consideration were found to be of primary importance. In the early 1920s some of the older steelmaking processes were being discarded; new ones were coming into use which required iron ores of different analyses and produced steels of superior quality with more accurate chemical control. Specifically, the decline in the use of the Bessemer process and the expanding use of the open hearth method of steelmaking materially reduced the need for ores low in phosphorus. At the same time, the increasing cost of coking coals made ores that were low in silica more desirable.

Mesabi Sinter had still another disadvantage. In the agglomeration of the taconite concentrate, it was necessary to make hard, glassy sinter that would withstand shipment without excessive breakage. This glassy product did not find favor with blast furnace operators because it was refractory and smelted slowly. They preferred softer, more completely oxidized ores. Perhaps these changes could not have been anticipated at the time the Mesabi Iron Company's programs were being established, but if men with more experience in the iron and steel business had been present in its management, serious questions regarding impending changes in the industry might have been raised.

The failure of the Babbitt taconite project was a sad blow to the people who had invested so much of their lives in the endeavor. While it represented the loss of a large sum of money to Jackling, Hayden, and most members of the board of directors, it meant something quite different to those who were closer to the actual operation. To Swart and Jordan and the people they had brought into the forlorn, burned-over country of the eastern Mesabi to build and operate the town of Babbitt, the mine, and the plant, it meant that the efforts of their minds and hands had been to no good purpose. Williams, Swart, and Jordan were the hardest hit. They were too far along in years to give taconite another try, and they had begun with such high hopes. Swart had said, "Selling our sixty per cent sinter will be just like selling fresh eggs in the winter." When the plant closed down, Fred Jordan remarked, "I'm going to get me a job digging ditches. You don't have to plan or worry, and when five o'clock comes you just throw down your shovel and go home."

Williams was very depressed. Although he did not have a

great deal of money invested in the project, he had put into it so much time and energy that the shutdown was a sad blow. He was a cool, analytical lawyer, and this was his first intimate contact with the failure of an operation. (St. Clair, on the other hand, bounced up optimistically and started immediately on another mining venture.)

Like Swart and the rest of us, Williams had taken a lot of ribbing from Duluth men in the iron mining industry who resented the Mesabi Iron Company. They regarded its activities as an effort on the part of Western copper men to show up the local talent. After the plant closed down, some people in Babbitt said openly that the "Duluth crowd" had somehow engineered the big drop in iron ore prices in 1924 just to "spank the copper crowd and send them back to Utah." There is nothing in the records to substantiate this idea. As always, the Lake Erie Base Price was established in 1924 by the price of early season sales; the year was a depressed one in the iron ore business. Total shipments from the Lake Superior district dropped from over 60,000,000 tons in 1923 to 43,896,000 in 1924. It was a coincidence that the price break occurred soon after plans to expand the Babbitt plant got under way. There were, however, men in Duluth who had predicted failure for the Mesabi Iron Company's taconite venture, and who seemed only too pleased that their predictions had come true.[30]

4 Back to the Laboratory

AFTER THE MESABI IRON COMPANY's plant at Babbitt closed in 1924, Minnesota taconite was very dead for a long time. Those of us on the staff of the University of Minnesota's Mines Experiment Station felt that the failure of this first commercial venture had set back the possible development of Minnesota taconite many years, if it had not killed it completely. Jackling and his engineers had a high reputation in technical as well as financial circles for their ability to overcome great difficulties in the development of low-grade ore properties. If this experienced and well-financed group could not make a success of taconite processing, who could? To many people it seemed a waste of time and money to try any more.

However, some informed people with whom we discussed the situation felt that one failure should not necessarily lead to abandonment of the entire idea.[1] It did not seem right to them to wait until all of Minnesota's higher-quality ores were gone before starting taconite utilization. The potential importance of taconite to the economic future of northern Minnesota made imperative at least a careful analysis of the cause of the Mesabi Iron Company's failure. Those of us in charge of the Mines Experiment Station — Dean Appleby, Henry H. Wade, chief metallurgist, and myself as superintendent — felt some responsibility for the project's collapse, and we set to work to find the cause.

All the technical results secured at Duluth and Babbitt were available to us. After analyzing them and giving the matter much serious study, we became convinced that the project was "ahead of its time" only in the sense that research had not developed processing methods to provide a pure enough concentrate or a superior agglomerate. We came to believe that, while technical difficulties had certainly contributed, the principal reason for the collapse of the Babbitt project was the failure to plan for the production and marketing of a more desirable, higher-grade product. The sinter made at Babbitt in the early 1920s was not a superior product but was barely the equivalent of Old Range ores. It was not in great demand then, and it would not be in great demand today.

We concluded that the Mines Experiment Station should not abandon taconite and that additional funds should be requested from the university for new equipment and increased staff to develop better methods of concentration and agglomeration. The objective of the station's research, we decided, must be the making of a superior product that could be used by the iron and steel producers to turn out better and cheaper steel than could be made from natural ores. We believed that it would be necessary in our new research efforts to forget the Lake Erie Base Price and the various premiums that might or might not be secured on a taconite product. Because ore processed from taconite would inevitably cost more than that simply dug out of the ground or beneficiated by standard methods, the success of any future taconite endeavor would, we felt, depend upon making a product that blast furnace operators could smelt more cheaply than natural ores.

Although I had been appointed superintendent of the Mines Experiment Station in 1918, the decision to go ahead with taconite research was, in the final analysis, made by Dean Appleby. He agreed to the program knowing that expenditures for such work would certainly be criticized by some people within the university and probably by some of the state's mining interests which had large investments in natural ores. He knew also that he would be called upon over the years to sponsor requests for additional funds from the Minnesota legislature to finance this expanded research.

The dean had already recognized the need for a better building and additional facilities for the Mines Experiment Station. In the spring of 1923 we moved into the brick structure which still houses the station on the university's Minneapolis campus, and during the next two years, much of the staff's time was consumed in constructing, purchasing, and installing the new machines needed.[2]

In 1925, after we decided to go ahead with taconite research, Dean Appleby and I appeared before the appropriations committees of the Minnesota legislature under the sponsorship of Edward P. Scallon, a mining executive and a state representative from Crosby. There, for the first time, we told the story of the state's declining high-grade iron ore resources and outlined the need for a program to assist and encourage industry in the utilization of low-grade ores, especially taconite. The legislature responded generously and for many years thereafter continued to provide appropriations, or "specials," as we called them. Although they had various formal names, they were always used for low-grade ore and taconite research. For the next twenty years these appropriations averaged about $22,000 a year. The station also received, for its general maintenance and support, additional sums of from $30,000 to $40,000 each year from the university's regular budget.

With a new building and increased funds, we at once began to expand the work of the station. Until 1921, Henry Wade, as chief metallurgist, James H. McCarthy, as chief assayer, and I had been the only full-time members of its technical staff. Mr. Wade had been graduated from the University of Minnesota School of Mines in 1915 and had immediately joined me at the experiment station. During all the years I was there, we worked closely together. As chief metallurgist, he was in direct charge of the operation and maintenance of the station's laboratory. In 1921 John J. Craig and Carl L. Wallfred joined the staff as assistant metallurgist and chemical engineer respectively, and in 1923, when the new building was ready, Charles V. Firth began his long and productive career with the organization. Bernard J. Larpenteur came in 1925; John C. Durfee, who became Mr. Wade's assistant, in 1928; Wayne E. Apuli in 1930;

Gilbert G. Willson in 1931; Harold H. Christoph in 1935; and Leslie S. Taylor in 1945. Many of these men were Minnesota graduates; they worked with Mr. Wade and me on taconite and related projects for many years. To this team belongs the credit for the development of the taconite concentration and pelletizing processes now in general use.

The work of the Mines Experiment Station in the 1920s and 1930s was divided into two principal branches, which we called "state service work" and "research." The former included examining, analyzing, and testing minerals submitted from any part of the state. It went on continuously, as it does to this day. Hundreds of samples are sent in by people from all parts of the state who wish to identify the material and find out whether or not it is valuable. In addition, many tons of iron and manganese ore were shipped to the station by large and small Minnesota mining companies for the purpose of determining whether or not the samples submitted could be beneficiated and, if so, by what means. If the minerals came from Minnesota, this work was done without charge.

The research carried on by the staff varied over the years as new projects were initiated and old ones were finished. For example, a study of oxidized or nonmagnetic taconite (now misnamed semitaconite) occupied our attention for some five years and resulted in the development of a magnetic roasting furnace and concentration plant which was erected at Cooley, Minnesota, in 1934. Other projects involved methods of producing manganese and iron powder from Minnesota ores. From 1918 until 1955, while I was in charge of the station, most of our investigations were related to low-grade ore — but not always taconite — in one way or another. Those mentioned are merely examples. After 1939, however, most of the other studies were set aside for later investigation, and we concentrated more and more upon the problems of magnetic taconite.[3]

Taconite Research Program

For a number of years after operations were discontinued at Babbitt in 1924, the large amount of publicity received by the Mesabi Iron Company's project caused widespread interest in the utilization of taconite-type materials found elsewhere. From the 1920s through the 1940s carload samples for testing were submitted to the Mines Experiment Station from Black River Falls, Wisconsin, from the Benson, Port Henry, and Clifton properties in New York, from Cornwall, Pennsylvania, and from Canada, Russia, Manchuria, and Spain.

This outside work brought to the station many technical experts, men with training and experience in particular operations and equipment.[4] By consulting these men and working on the samples of non-Minnesota ores, we learned much about the production of high-grade concentrate from ores requiring fine grinding. The magnetic ore shipped to us from New York, for example, was far simpler to concentrate than Minnesota taconite, since it contained over 40 per cent iron, was coarsely crystalline, comparatively soft, and easily crushed. In spite of these differences, the tests we made on it furnished us with valuable information on the amount of power required to grind taconite in large commercial equipment. Tests on other ores also enabled us to develop certain ideas that could be used in magnetic taconite concentration flow sheets, although at that time we had no east Mesabi taconite to work with at the experiment station. All in all, the knowledge we gained and the new equipment we acquired for this outside work later proved very helpful in our research on Minnesota taconite.

At various times as opportunities arose, we consulted Eastern blast furnace operators and steel plant engineers to learn what they would consider the most desirable type of ore for smelting.[5] Although there was no close agreement among these men, they all mentioned certain requirements. (1) They wanted the ore to be low in sulphur, phosphorus, and earthy impurities which produced undesirable or excessive amounts of slag. (2) They preferred the iron to be in the form of Fe_2O_3, hematite, rather than Fe_3O_4, magnetite, because the former was believed to reduce more rapidly in the smelting furnaces. (3) They noted that the ore should be physically high in microporosity. (4) They agreed that the ore pieces should all be about the same size, and, strangely enough, of a rounded shape, although this seemed to be an unpremeditated selection. On the size of the pieces, however, the men had various opinions. One wanted them the size of baseballs, while another thought they should resemble grains of wheat.[6]

To us at the Mines Experiment Station, these preferences indicated that a desirable taconite concentrate should contain about 8.5 per cent or a little less silica (which means about 64 to 65 per cent iron), that the fine concentrate should be agglomerated by some means to change the magnetite to hematite, and that the final hematite product should be in the form of rounded lumps of uniform size and shape, highly porous and well oxidized. With these requirements in mind, we set to work in 1939 to perfect a processing method that would turn out a more desirable taconite product which would be both chemically and physically superior to the sinter that had been made by the Mesabi Iron Company.

For many years after the disastrous failure in 1924, the Mesabi company's property at Babbitt stood idle. The firm had to borrow money to maintain its investment and to pay taxes. Funds were advanced by Jackling, Hayden, and other wealthy members of the board of directors, but interest became weaker, and it seemed just a matter of time until the firm folded up entirely. In 1937, however, Jackling and Rolland C. Allen, vice-president of Oglebay Norton and Company of Cleveland, became associated as president and president-elect of the American Institute of Mining and Metallurgical Engineers. Allen was interested in large ore reserves, and by a series of steps, Oglebay Norton, acting as agent for four steel firms, organized the Reserve Mining Company in 1939. Reserve then leased the Mesabi Iron Company's property near Babbitt.[7]

The Mines Experiment Station's first contact with the new company occurred when Mr. Wade met Verne D. Johnston, a geological engineer employed by Oglebay Norton, and the two men discussed the Babbitt property. I then wrote Johnston on October 3, 1939, explaining that we had in the past done a great deal of work on Babbitt ore, that 64 per cent concentrate could be produced from it, and that we had worked out a flow sheet to accomplish this. Johnston later came to visit the station, and in 1941 samples of taconite from the Babbitt property were shipped to us. This was the first east Mesabi taconite rock we had received since the Mesabi Iron Company's failure seventeen years before, although we had obtained concentrate and fine sinter from storage.[8]

As it worked out, our second round of research on Babbitt taconite was divided into two general projects. The first involved fine grinding and magnetic concentration to produce at least a 64 per cent iron product. The second had to do with the agglomeration of these fine particles of concentrate into the rounded, well-oxidized lumps that the blast furnace operators had specified. The processing of taconite is simple in theory but complex in execution. The extremely hard rock must be crushed and ground to a fineness resembling flour. This fine grinding liberates the small particles of high-grade magnetite. These are caught and removed (or concentrated) by magnetic separation. The particles must then be put back together (agglomerated or pelletized) to make pieces large and hard enough for shipping and smelting. The problem of fine grinding and magnetic concentration was assigned to Henry Wade, who was assisted by John Durfee and later by Harold Christoph. Their objectives were to develop a flow sheet and the necessary equipment to produce the high-grade concentrate. John Craig and Charles Firth took over the task of developing an agglomerating process that could be used to make a superior blast furnace ore out of the fine magnetic concentrate that Mr. Wade and Mr. Christoph produced. These two projects more or less interlocked, and often both groups, which were very informal and co-operative, worked on the same problem.

The Taconite Concentration Investigation

In the processing operation required to produce the 64 per cent iron concentrate, Mr. Wade and his group recognized that three steps had to be carried out in proper sequence under controlled conditions: (1) The ore from the mine must be crushed into small pieces; (2) These pieces must be ground fine enough to pass through a 325 mesh screen (about the fineness of flour) in order to produce the desired grade of concentrate; (3) During or after the size-reducing operation, two-thirds of the rock must be rejected as low-grade nonmagnetic tailings.

Specially designed machines were available for these operations, but each machine had its particular field of usefulness and its own operating requirements. The problem before Mr. Wade and his group was to prepare a flow sheet, first on paper and then using equipment in the laboratory, that would satisfy the requirements of each machine and that would most efficiently produce the desired results.

To accomplish the first step of crushing the ore from the mine into manageable pieces, two types of commercial crushers were considered satisfactory: gyratory and cone crushers. Crushing is a general term that may apply to the breaking of pieces of rock of any size, but for taconite it generally refers to the reduction of large pieces of ore, perhaps as big as a piano, to about three-fourths of an inch. Size reduction below this point is usually called grinding. For the primary crushing of the three- or four-foot chunks of taconite from the mine, gyratory crushers were best. These enormous machines, weighing several hundred tons, could break great blocks of taconite to about twelve inches in size at the rate of a thousand tons or more an hour. Smaller gyratories could then reduce the twelve-inch chunks to about three inches; from that size, cone crushers could be used to break them into three-quarter-inch pieces.

During the final coarse-crushing stage — which might require two or more cone crushers in series — it is desirable to remove the pieces already crushed finely enough. This is done by means of screens which must be designed to resist the wear of the abrasive taconite. They may be heavy woven wire of the required mesh opening, steel plates with holes of the desired size, or simply parallel steel bars spaced the desired distance apart. The screen surface slopes at a small angle and vibrates so that the ore slides slowly over it. The pieces smaller than three-fourths of an inch fall through the openings, and the larger pieces go to the final cone crushers.

Such powerful equipment was not, of course, available in the Mines Experiment Station's laboratory, but Mr. Wade and his men observed it in satisfactory operation in commercial plants crushing hard rock. In the laboratory, we had crushing and screening equipment to handle six- or eight-inch pieces of taconite; we broke the larger ones by hand with heavy sledges.

The next stages of the process—grinding from three-fourths of an inch to 10 mesh and then to 325 mesh—could be carried out in two steps. At Babbitt a series of unsatisfactory crushing rollers had been used for the first portion of this work. Since the Babbitt plant closed down, a new machine called a "rod mill" had been developed and was being used effectively on the softer copper ores in the West. In 1942 we decided to secure one of these machines and try it on our hard taconite. It proved to be very satisfactory for grinding three-quarter-inch taconite to 10 mesh, or about the consistency of coarse sand, and it greatly simplified the first-stage grinding problem for Mr. Wade and his group.

A rod mill is a large, heavy, cylindrical steel container lying on its side and mounted to rotate on hollow bearings, or trunnions. Ore with water is fed into one end of the rotating container, and a mixture of crushed ore and water (pulp) flows out through the hollow trunnion at the other end. Our laboratory rod mill was three feet in diameter and six feet long, but commercial mills may be ten or twelve feet in diameter and sixteen to twenty feet in length. They are lined with thick steel wearing plates, and they rotate slowly (normally from fifteen to twenty times a minute). The larger the mill, the slower the speed.

These mills are normally about half full of steel rods nearly as long as the inside of the mill. For grinding hard rock from three-fourths inch to one-eighth inch as we hoped to do, we found that rods as large as four inches in diameter were being used. These rods, weighing about 500 pounds each, are not attached to the mill or to each other; they are free to roll upon one another and the inside surface of the mill, grinding the ore particles that are caught between them. The rods wear with use and gradually become smaller. New large rods are added frequently, with the result that the mill is constantly about half full of rods ranging in size from four inches or so down to about one inch in diameter. At the latter size they break up into short lengths and are discharged with the ore.

For the last fine-grinding step — from 10 mesh to 325 mesh, or from the consistency of coarse sand to that of flour — Mr. Wade's group found that a ball mill was the most satisfactory machine. Like a rod mill, it is a cylindrical container mounted to rotate on hollow trunnions. In outward appearance the two machines look much alike. Both are about the same size, and both rotate at about the same slow speed. Inside, however, they are quite different. The ball mill is about half filled with steel balls two or three inches in diameter. As the mill rotates, the balls roll upon one another and the mill lining, just as the rods do in a rod mill. Like the rods, the balls wear with use and gradually become smaller. New balls are added frequently, and the worn balls, a half inch or less in diameter, are discharged with the ore.

Determining the proper size of the balls or rods to be used in these mills was very difficult. Mr. Wade and his men worked out the requirements for east Mesabi magnetic taconite only after many daylong tests. They were guided by the results of similar grinding operations in various plants and their own experience at the station with other ores, none of which was as hard and abrasive as taconite.

For east Mesabi taconite, they found that the grinding characteristics of the rod and ball mills were quite different. The rod mill was effective in grinding the three-quarter-inch ore to 10 mesh, but below that size grinding seemed to taper off rapidly. The ball mill, on the other hand, was rather ineffective on ore particles larger than 4 mesh, but was quite effective in grinding 10 mesh and smaller particles to 325 mesh and finer.

Many particles, however, passed through the ball mill without being ground as fine as desired. To solve this problem, those that were fine enough (325 mesh) were removed, and the coarser particles were returned to the same ball mill to be reground. The machine in general use to separate the coarse particles from the fine ones is called a hydraulic classifier. It works on the principle that coarse particles settle more rapidly in water than fine ones. When the ball mill product is put into a container with a large amount of water, the finer particles tend to overflow with the water and the coarser ones settle to the bottom, where they can be withdrawn for regrinding. The classifier is not as efficient as a screen in making a size separation, but screening hard, abrasive taconite at sizes below one-fourth inch, let alone 325 mesh, is much too difficult and costly.[9]

Perhaps a better understanding of the crushing and grinding problem can be secured by considering the number of pieces to

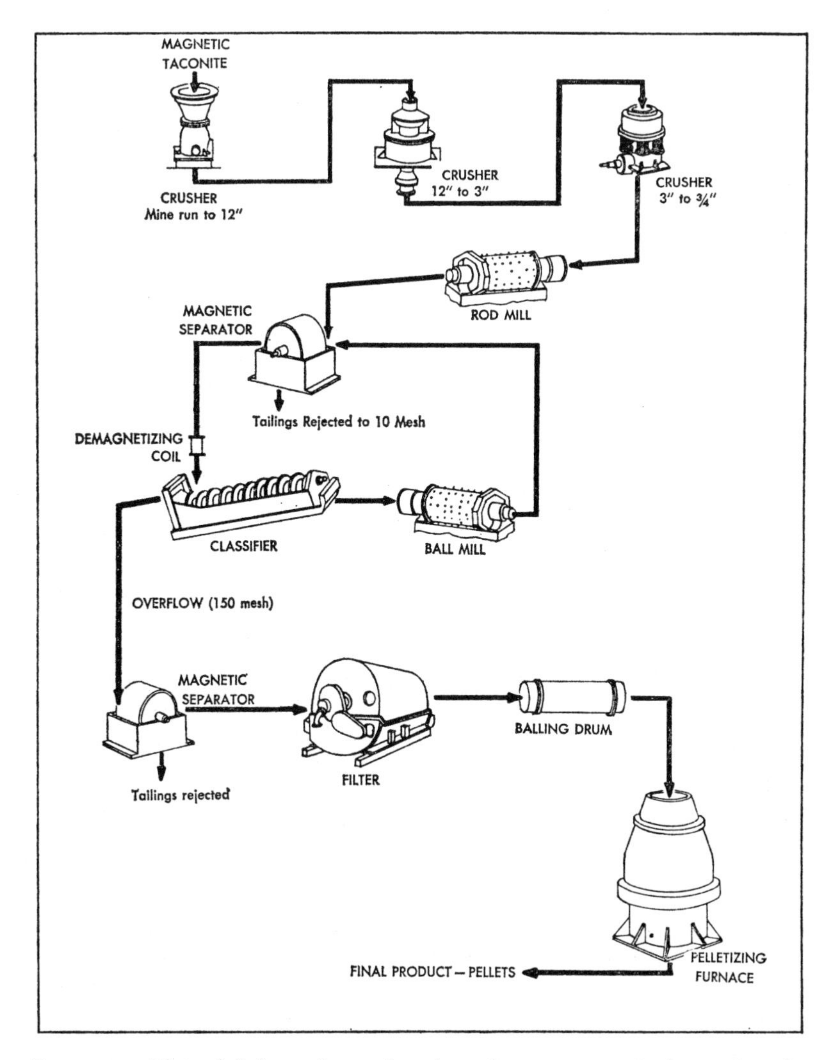

FIGURE 7. *Pictorial flow sheet showing the steps worked out by the Mines Experiment Station for processing magnetic taconite. The station's experiments started with the rod mill.*

be broken at each step. Energy is wasted if harder or lighter blows are struck than are needed to break the piece. For example, it is a waste of energy to use a sledge if a tack hammer will do the job and vice versa. A cube of taconite measuring three feet on each side weighs nearly three tons. When it is dropped into a big gyratory crusher, it is caught between the jaws and broken, let us say, into 27 one-foot cubes. In the next crusher, each of these 27 pieces is broken into 64 pieces three inches in size. The second crusher does not have to strike such heavy blows, but it must strike 27 of them and in doing so produces a total of 1,728 pieces. The third crusher then strikes and breaks 1,728 three-inch pieces to three-fourths of an inch in size. Again lighter blows are needed, but 1,728 of them are required, thus producing 110,592 three-quarter-inch pieces from a single three-foot block of taconite weighing about three tons.

These are fed to the rod mill which breaks each piece to one-eighth inch, producing about 24,000,000 pieces to be ground about as fine as flour in the ball mill. This mill produces nearly ten trillion 325 mesh particles from the original three-foot block of taconite. Since taconite does not break into cubes and since in each crushing and grinding step pieces of all sizes are produced, the above figures are, of course, only approximate. They do, however, give some indication of the crushing and grinding problem in each size-reduction step.

Each succeeding crushing step becomes more expensive per ton of taconite ground. The power required by a ball mill to grind a ton of taconite from 10 mesh to 325 mesh is greater than that needed for all the previous machines put together. Realizing this, Mr. Wade and his men made every effort to reduce the work required of the ball mill by discarding the pieces of taconite containing little or no magnetic iron before they reached this mill. However, we knew from earlier experiments and from experience at Babbitt that the finer the ore was crushed before tailings elimination, the greater the amount that could be rejected by the magnetic separators at any particular iron assay. It was therefore necessary to balance the value of the iron lost in the tailings against the cost of additional crushing and grinding in order to arrive at the best over-all efficiency for the whole operation. This involved many tests and voluminous calculations.[10]

After fine crushing, particles of taconite can be divided under a microscope into four recognizable types, as shown in Figure 8. Particle A is high-grade magnetite and should be included in the concentrate. Particle D is silica and should be discarded as tailings. Particles B and C show two types of middlings, the name given to particles that are a mixture of magnetite and silica. Both contain too much magnetite to be discarded and too much silica to be called concentrate. Both also contain sufficient magnetite to be attracted by the magnets in the magnetic separators. It is apparent that if Particle B is crushed finer—along the dotted lines, for example—pieces will be produced that contain sufficient magnetite to be included with the A particles as concentrate, while the remainder can be discarded as tailings.[11]

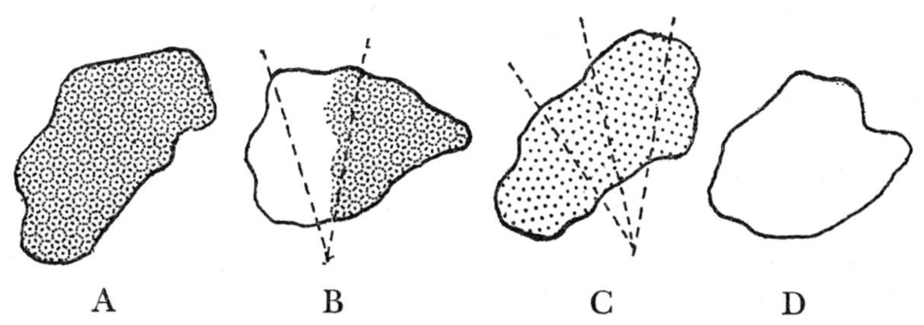

FIGURE 8. *Typical 325 mesh particles of taconite greatly magnified. Ten such particles could be laid on the period at the end of this sentence without overlapping.*

The C particle presents a different problem. It is made up of what are called "dusty magnetite" particles uniformly dispersed in the silica. Crushing along the dotted lines breaks the particle into smaller pieces, none of which contains a greater percentage of magnetite than the original. If the amount of this type of middlings is not too great, it can be absorbed by the concentrate without seriously affecting its assay. But if the amount is large, it may not be possible to produce low-silica concentrate without discarding these middlings and causing a severe loss of iron.

Moreover, with particles of the C type, magnetic concentration is not effective even after the finest commercial grinding.[12]

The dusty middlings particles may contain as much as 20 or 30 per cent magnetic iron, which is sufficient for them to be strongly attracted by the magnets. If so, they will enter the concentrate and thus increase the silica content above the desired limit. It may, therefore, be necessary to remove this type of middlings from the concentrate by some method other than magnetic separation, such as careful hydraulic classification or, more effectively, by flotation.

Flotation is a concentration method used extensively on copper but also applicable to iron ores. In this process the finely crushed ore in water is mixed with small amounts of various carefully selected chemical reagents like soap or oil. When this mixture is strongly agitated, a thick scum or froth forms on the surface. If this is carefully skimmed off, it will contain most of the silica or most of the iron mineral in the ore, depending upon the reagents used. Flotation is a much more expensive process than magnetic separation, but it is more effective for the removal of troublesome middlings particles.

Side magnetic effects also can cause trouble in magnetic concentration. In strong magnetic fields, the particles of magnetite become small permanent magnets, and they tend to stick together even after they are removed from the magnetic separator. If this occurs, middlings particles similar to B get into the concentrate and can be disentangled only with difficulty. Most magnetic separators are designed to agitate vigorously the concentrate by magnetic and hydraulic means in order to liberate as many of these low-grade ore particles as possible.

It should be understood that the middlings problem is not as simple as the above illustrations would seem to indicate. Actually, there are all conceivable combinations of the various particles shown. Few tailings particles of east Mesabi taconite — even as small as 325 mesh — can be selected that contain absolutely no magnetite; conversely, few particles of magnetite can be separated that contain absolutely no silica. The amount of dusty magnetite middlings in Mesabi taconite varies widely from place to place and from one horizon to another. In general, the slaty horizons contain more such particles and the cherty horizons contain less. Metallurgists classify taconite as having low or high "concentratability," which means principally having a large or small amount of this type of middlings. From the standpoint of concentrating engineers, the ease or difficulty of processing taconite from any particular locality or horizon depends upon how completely fine grinding liberates the magnetite from attached or included particles of silica.

The magnetic concentrators available in the Mines Experiment Station's laboratory consisted of the tube machine which had been mechanized and standardized; a dry cobber, for which we had little use; a small magnetic log washer of the type used at Babbitt; the original wet cobber from the Duluth experimental plant; and a three-drum counterflow magnetic separator from Bethlehem Steel Corporation's Lebanon plant. The latter was similar to the wet cobber previously described in Chapter Two, but it had an arrangement of baffles to direct the flow of water counter to the movement of the ore and thus increase the washing action.

The wet cobber had worked well in the Babbitt plant on ore crushed to about 4 mesh, but it had been undependable because water got into the windings of the electromagnets and grounded them. Mr. Wade eliminated this difficulty simply by immersing the magnets in transformer oil. This improvement turned the wet cobber into a reliable machine, and hundreds of them are now in general use in magnetic plants the world over.

With the new rod mill, a ball mill, a reliable wet cobber, a modified Lebanon magnetic drum separator, and other auxiliary equipment such as classifiers, pumps, and dewatering tanks, Mr. Wade and his group were in a position to prepare flow sheets and arrange equipment in the laboratory to test them. No attempt was made to set up a flow sheet of equipment for the coarse crushing of taconite. Our experiments started with the rod mill, and all the steps involved wet processing.

As finally assembled the new Mines Experiment Station and the old Babbitt flow sheets look much alike when placed side by side, but they differ in three details that greatly affect the efficiency and economy of the whole operation. (See Figure 9.) The first is the use of a rod mill in place of the crushing rolls; the second is the elimination of all dry cobbing; and the third is

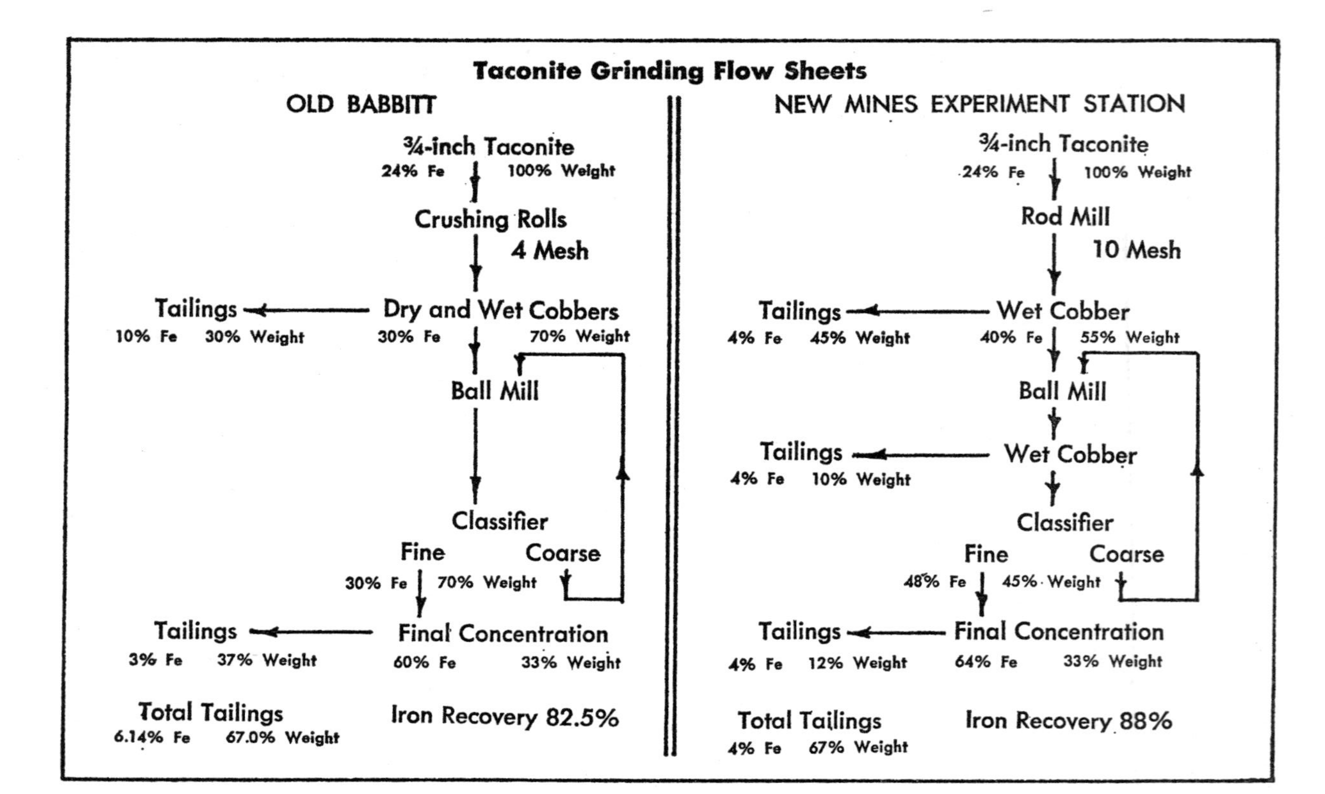

FIGURE 9. *A comparison of the Mesabi Iron Company and the later Mines Experiment Station flow sheets.*

the addition of a wet cobber following the ball mill. The development of these machines and their proper use made the new flow sheet greatly superior to the old one.

With the rod mill we could crush the taconite much finer than we could in the rolls, thus making possible the elimination of more and lower-grade tailings by the first wet cobber. This resulted in a richer, finer feed for the ball mill to grind and much less of it, which materially reduced the work of this mill.

In the Babbitt flow sheet, the ball mill discharged into a classifier and the coarse product was returned to the ball mill for further grinding (closed circuit). Any ore fed to the ball mill could escape from the closed circuit only after it had been ground fine enough to be discharged by the classifier, regardless of whether it was magnetite or silica. Adding a wet cobber to the ball mill circuit changed this. As soon as a piece of ore was broken, the liberated particles of silica were discharged as cobber tailings. With the new flow sheet, the ore fed to the ball mill could escape from the grinding circuit either as coarse tailings or as a fine classifier product, thus further reducing the work of the ball mill.

From a comparison of the weights and assays shown on the two flow sheets, it is evident that the new one retained a higher percentage of the iron content of the taconite and more efficiently rejected the tailings. In the old flow sheet, only 30 per cent of the original weight of the rock had been eliminated before reaching the ball mill; therefore this mill had to grind 70 per cent of the ore fine enough to overflow the classifier. In the new process, 45 per cent of the original weight had been eliminated as tailings before it reached the ball mill; 10 per cent more was rejected by the wet cobber in the ball mill grinding circuit, leaving only 45 per cent of the ore to be ground fine enough to overflow the classifier. These changes in the process not only materially increased the capacity of the grinding circuit, but also raised the grade of the final concentrate from about 60 to 64 per cent iron with no loss in the weight of the concentrate recovered (33 per cent). The recovery of iron in the concentrate was 88 per cent compared to 82.5 per cent in the old Babbitt flow sheet.

The equipment in the experiment station's laboratory could be arranged in the new flow sheet for continuous operation at a feed rate of about one and a half tons of three-quarter-inch taconite to the rod mill per hour. From the 10 mesh rod mill pulp, the first wet cobber rejected almost half of the ore as low-grade tailings, thus reducing the bulk to a little over three-fourths of a ton per hour to the ball mill. The second cobber (in the ball mill grinding circuit) still further reduced this to about five-eighths of a ton. Thus the final concentration equipment, consisting of magnetic separators and washing tanks, had only one-eighth of a ton of additional tailings to remove in order to produce the final product of a half ton of concentrate of the desired grade of 64 per cent iron. Mr. Wade and his men made literally hundreds of tests with this equipment adjusted to operate in various ways. The final result was the most satisfactory processing method yet devised to produce 64 per cent iron concentrate from east Mesabi taconite containing only a comparatively small amount of dusty middlings.

Agglomeration Investigation — Pelletizing

The fine concentrate made by Mr. Wade and his group from several carloads of taconite shipped to the university over the years by Frank Emanuelson and his crew at Babbitt was turned over to Craig, Firth, and Taylor. Craig worked on the problems of agglomeration off and on from 1925 until he left the Mines Experiment Station in 1942; Firth carried on the work until his death in 1945; and from that time until after my retirement in 1955, Taylor continued the study. Several hundred tons of concentrate produced in the laboratory — a half ton at a time — were turned over to these men, who conducted a series of lengthy experiments, with numerous interruptions and distractions, over a period of thirty years (1925–55) to develop a better method of agglomeration that came to be known as pelletizing.

With the special appropriations for low-grade ore research received from the legislature beginning in 1925, John Craig began an investigation of sintering. At the time, sintering was the standard process for agglomerating fine ore. It was not a cheap operation, requiring as it did for taconite concentrate about

140 pounds of coal or coke per ton of product. Moreover, with material as fine as this concentrate the capacity of a sintering machine was low, and the hard, glassy sinter was not especially desirable for blast furnace smelting. Its glassy structure and low microporosity made it rather impervious to the furnace gases, and it reduced slowly.

Since the Mesabi Iron Company had produced and sold a sintered product, however, it seemed important to learn more about the fundamentals of this process. In order to separate fact from theory, John Craig built sintering equipment of several kinds and experimented with it. Professor George M. Schwartz of the university's department of geology worked with us, studying petrographically the samples of sinter which we procured or made under various conditions using different kinds of ore. He found that sinter made from taconite concentrate was composed largely, but not entirely, of small grains of magnetite cemented together by fayalite, an iron silicate (or slag) which has a low melting point. He told us that under certain conditions there was also considerable grain growth (large crystals forming at the expense of smaller ones), and that magnetite crystals were often converted to hematite.[13]

All the samples of hard, commercial sinter we examined showed that iron silicate had formed and melted. The generally accepted notion among sintering plant operators at the time was that it was the fusion of the iron silicates which made the sinter hard and strong. If this were true, pure iron oxide would not make good sinter, and the ore to be sintered must contain a certain amount of silica in order to form the low-melting-point iron silicate. To check this theory, we removed all the silica from a sample of taconite concentrate and mixed in pure carbon as fuel. We found that good sinter could be made from this material, although no iron silicate was formed. At the other extreme, we tried sintering pure silica and learned that this, too, could be done. Thus we concluded that the presence or absence of silica in the ore to be agglomerated was not a controlling factor.

At the time these tests were made, it was also believed that only fine anthracite coal or coke could be used as fuel for sintering. Another series of experiments showed, however, that under proper conditions, good sinter could be produced with soft coal, charcoal, lignite, sawdust, and even peat. By these investigations we cleared up many of the mysteries of sintering that had plagued us in the Babbitt plant. Through them we learned that the only requirements for good sinter were sufficient fuel and air to produce a localized high temperature, and that just as good or better sinter could be made by interlocking grain growth as by the cementing action of low-melting-point iron silicate. The information thus gained was later applied to the taconite pelletizing process we eventually developed.

In another investigation carried on in the 1920s and 1930s we tried to convert fine taconite concentrate into metallic iron without the agglomerating step required for blast furnace smelting. In the copper industry, as the better natural ores were exhausted and it became necessary to utilize leaner ores, a new type of furnace, called a reverberatory furnace, was designed that would smelt fine concentrate. Reverberatory smelting was not a direct reduction process since the final product was in the form of molten iron. If it proved successful for taconite, however, it might, like direct reduction, eliminate the need for a blast furnace.[14]

Using a reverberatory furnace on fine taconite concentrate, we were successful in producing a type of low-quality pig iron, which we called "crude iron." The trouble was no one wanted this type of metal, because it contained too much sulphur for steelmaking. Then we learned that cast iron was being used to some extent in Europe as a paving material for streets on which horse-drawn traffic caused severe wear. It seemed to us that our crude iron might be useful for this purpose, and that a large local market might be developed for the pig iron produced by the reverberatory smelting process.

We produced what seemed to be a superior paving block — six inches square, about two inches thick, and with a patterned tread surface not unlike that of a modern auto tire. In 1934, as an experiment, two sections of pavement using these blocks were laid on Fifteenth Avenue on the university campus; three years later a half-mile stretch was paved with them near Eveleth on the Mesabi Range. The blocks proved to be about as satisfactory a paving material as brick, but they cost considerably more to make, transport, and lay. Moreover, they were too noisy for high-speed auto traffic, and they were not as skidproof as modern transportation required. The effort turned out to be just another of those "good" ideas for the horse and buggy days, and we returned to our original idea of agglomerating the taconite concentrate for standard blast furnace smelting.

After our conferences with blast furnace operators in the late 1930s, Craig and Firth studied the literature and all the patents they could find on the agglomeration of fine iron ore. Their research showed that many techniques had been used, but the one which produced the agglomerate that came closest to meeting the specifications of the blast furnace experts was the Gröndal briquetting process developed in Sweden in the early 1900s by Gustav Gröndal.[15] I had seen it in operation in 1917 at Moose Mountain and had returned with some of the briquets.

In briquetting processes, the fine damp ore, with or without a binder added, is pressed or compacted into masses. These are then dried, heated, or actually fired, depending upon the binder used or other requirements of the process. Briquets of many sizes and shapes have been made, using such binders as lime, cement, and tar. In other types of briquets (pellets, for example) no binder is required; the particles of fine ore are held together by molecular and crystalline intergrowth (as are many ordinary rocks). Only this type of briquet or pellet is satisfactory for blast furnace smelting.

The Gröndal process was not unlike that used in modern brickmaking, and the final product was a hard, burned block about the size and shape of a modern building brick. Although the briquets were not the size or shape our blast furnace operators preferred, they did meet several other requirements that these men had mentioned. The briquets were hard, highly porous, and strong enough to withstand handling, and the magnetite had largely been converted to hematite. Those we examined seemed to have been of ideal structure for blast furnace smelting, but the process had been generally abandoned in the intervening years because it was so expensive, involving considerable hand labor and large amounts of fuel. Sintering had dominated the field of iron ore agglomeration since the 1920s.

We undertook to make a product from fine taconite concentrate having about the same structure and analysis as the Gröndal briquets but of a different shape. Like the Gröndal process, our operation consisted of two steps: compacting the damp, fine concentrate, and then firing the compacted pieces or "compacts," as they were called, below fusion temperature. It soon developed that, while the firing of the compacts would undoubtedly be the most expensive step because of the fuel required, the making of the compacts would take the most study. The whole operation was not unlike the manufacture of pottery. Much skill and experience is required to properly temper the clay and form it into the desired shape, but once this is done, the firing is a comparatively simple operation that can be made practically automatic.

At first we attempted to make taconite compacts using an extrusion method. The cylinders of concentrate thus formed were one or two inches long and a half inch in diameter. These could be fired to produce substantial agglomerates. The extrusion equipment was awkward and expensive to operate, however, and we tried various other ways of making compacts, including a vibration system that seemed for a while to offer some promise.[16]

By 1938 we had learned that if balls of the damp concentrate were simply rolled by hand like small snowballs and then fired, hard agglomerates could be made that seemed to completely satisfy the preferences expressed by the blast furnace operators we had consulted. Although their desire for ore in a rounded shape seemed not to have been a premeditated choice, there was really a very sound theoretical reason for it. Spheres of equal size when placed in a container will leave about 35 per cent void space, which is equally distributed throughout the mass regardless of how the spheres arrange themselves. In the blast furnace, this means that more uniform gas flow and more rapid smelting should theoretically be obtained by using ore in the form of equal-sized spheres. The size of the spheres would determine the pressure required to force the air through the furnace charge, but the percentage of void space would remain about the same if the spheres were all the same size—whether they were large or small — thus giving the hot furnace gases equal access to all particles of the ore.

With this in mind, Craig began in 1938 a series of experiments to make firm, well-compacted balls from dampened taconite concentrate by some mechanical method that would lend itself to large-scale commercial processing. Again the earlier research of others provided a lead. In 1917 on a trip to New Jersey with Walter Swart, I had observed Henry J. Stehli and John E. Greenawalt experimenting with a concrete mixer that formed damp blast furnace flue dust into roughly rounded masses for sintering.[17] We decided to try to refine this operation and apply it to taconite concentrate. Craig secured from a junk dealer a section of an old smokestack about thirty inches in diameter and five feet long. He mounted it so that we could rotate it at variable speeds and at the same time change its slope. With the drum rotating and placed at a slight angle, we fed the damp concentrate into the upper end, and occasionally a few good, solid, round balls of about the desired size (along with unballed concentrate) rolled out the lower end. It was a very delicate operation, however, and often no balls at all came out. At other times, the whole mass of ore in the drum collected into a few large balls the size of grapefruit or footballs.

Our experiments to find out how to operate this drum in order to produce the desired size and quality of balls continued into the early 1940s. We learned to our frustration that if the speed of the drum was too high, the balls were thrown about and broken. If the slope of the drum was too flat, the balls varied greatly in size. If the ore was too wet, the balls formed rapidly, but they were too soft. If the ore was too dry, the balls formed slowly and they were too brittle to withstand handling. And the tendency of the ore to stick to the inner surface of the drum was always a difficulty.

On the other hand, we were encouraged to find that when we did manage to make a few good well-compacted balls, they were surprisingly strong. They could be dropped from a height of three feet to a hard floor without breaking, and they were so tightly compacted that they could not be made much stronger, even in a high-pressure press.

Eventually we secured the best results by screening out the balls that came from the drum at the desired size — usually about three-fourths of an inch — and then feeding the smaller balls and the unballed concentrate back into the drum with the new feed to be rolled again until balls had been formed that were large enough to be removed by the screen. Two or three passes through the drum might be required before the balls were of the desired size. In fact, we found that better balls were produced when the quantity of small balls being returned was large. If balls of the desired size were made in one pass through the drum, they were usually soft and easily broken. After hundreds of experiments, we established operating conditions that usually — but not always — produced satisfactory results. The whole balling operation was very baffling and seemingly temperamental. Slight changes in the nature of the feed and in its moisture content drastically altered the balling quality of the concentrate.

In large-scale tests, the balling drums we designed developed certain weaknesses, the most serious of which was a tendency for the damp taconite concentrate to stick to the inside of the drum. Often lumps several inches in thickness were formed, making the drum's interior rough and irregular. This condition interfered with the rolling action and caused the formation of large, irregular masses that fell off and made soft oversized balls. We worked off and on for nine years to find a method by which the inside of the drum could be kept smooth and uniform. We tried lining it with various materials, and for a time we thought that synthetic rubber was the answer. The problem was not really solved, however, until 1949 when Henry Wade installed a sort of boring bar equipped with hard steel cutting teeth that moved continuously back and forth near the inner surface of our small drum. Since that time, rotary drums of various sizes and shapes — some cylindrical, some conical, and some saucer-shaped — have been developed, but the movable leveling bar is now used on all of them.

While we struggled to perfect the balling operation, various members of the Mines Experiment Station's staff were working to find feasible ways of firing the balls to change the magnetite to hematite and to make pellets hard enough for shipping. After considerable experimentation in the late 1930s with small ones, we built at the station in 1940–41 a pelletizing furnace eleven feet high, with a rectangular shaft forty-eight inches long and twenty inches wide. It had combustion chambers on each side, and from these the products of combustion, at a temperature of about 1900 degrees Fahrenheit,[18] passed into the shaft through openings in the brickwork. When the shaft was filled with well-balled concentrate from the balling drum, the hot gases from the combustion chambers flowed upward, heating the balls to the desired temperature and escaping from the top of the shaft as relatively cool moisture-laden gases. Any kind of fuel desired could be burned in the combustion chambers, but the total required to produce about a ton and a half of fired pellets per hour amounted to approximately 3,000 cubic feet of natural gas or its equivalent.[19]

Over the years, we operated this furnace in many ways. Late in the 1940s we made a study of adding pulverized coal with the concentrate before the green balls, as they were soon being called, were formed. This additional fuel materially reduced the amount of gas required in the combustion chambers. Air, blown into the bottom of the shaft, flowed upward, cooling the balls to some extent and becoming hot enough to ignite and burn the coal contained in them. Later we found that the furnace could be operated without combustion chambers and without coal mixed with the ore simply by blowing a dilute mixture of gas and air directly into the bottom of the furnace shaft.

Most of the pellets we made were three-fourths of an inch in diameter, but we also tested green and fired pellets from one-eighth of an inch up to two inches in diameter. To simulate the weather conditions of northern Minnesota, the fired pellets were moistened and frozen, then thawed and frozen again many times. They were tested for strength and resistance to abrasion, both hot and cold, and for porosity and reducibility. Dr. Strathmore R. B. Cooke, professor of metallurgy in the university's School of Mines, made careful microscopic studies of the internal structure of the fired pellets, gave us the explanation for the strong bonding of the particles, and told us how they could be made still stronger.[20] Gradually, certain fundamental requirements for pelletizing became apparent, the two most important of which were: (1) good fired pellets cannot be made out of soft, misshapen green balls, and (2) taconite pellets should be fired at a temperature of about 2300 degrees Fahrenheit, using an excess of air to produce strongly oxidizing conditions.

The first public demonstration of the rectangular pelletizing furnace was conducted on July 22 and 23, 1942, using pellets of indifferent quality made in our still-temperamental balling drum. On April 22 to 24 of the following year we held our first continuous pelletizing demonstration. This time we used concentrate furnished by Percy L. Steffensen, director of raw materials research for the Bethlehem Steel Corporation, who had been experimenting with the sintering of Cornwall ore in Pennsylvania. Many mining company and other observers were present, representing not only Bethlehem but also such other firms as Algoma Steel Corporation Limited, Allis-Chalmers Manufacturing Company, American Cyanamid Company, Arthur G. McKee and Company, Butler Brothers, Cleveland-Cliffs Iron Company, Dings Magnetic Separator Company, Inland Steel Company, Jones and Laughlin Steel Corporation, Oglebay Norton and Company, and Pickands Mather and Company. A year later, in 1944, Charles Firth presented a formal paper describing the pelletizing process before the Blast Furnace Raw Materials Committee of the American Institute of Mining and Metallurgical Engineers.[21]

For over fifteen years with the support of the university and the legislature, the Mines Experiment Station, working alone, had made slow but steady progress in the development of taconite processing. Besides directing and supervising the work of the station, much of my time during that period went into keeping the taconite project alive. Every two years it was necessary to secure the funds required to support the expanded research staff at the station. This involved persuading the university to put our requests for funds into the budget which was presented to the legislature every two years, and then persuad-

THE MINES EXPERIMENT STATION'S BUILDING *on the Minneapolis campus of the University of Minnesota soon after its completion in 1923. (Rear view.)*

THE MAGNETIC ROASTING PLANT *at Cooley, Minnesota, processed semitaconite in the 1930s as an experimental unit under the direction of the Mines Experiment Station. The plant was then taken over by Butler Brothers. It closed in 1938.*

HENRY H. WADE *(left), John J. Craig, and Charles V. Firth of the Mines Experiment Station worked for many years on taconite processing.*

TYPICAL SAMPLES OF TACONITE AGGLOMERATES. *The sinter and nodules at left were produced in the 1950s by United States Steel's Extaca plant; the three-fourths-inch pellets by Erie's Hoyt Lakes plant. At lower right is a Gröndal briquet made at Moose Mountain in 1918.*

AN EXPERIMENTAL *Dwight and Lloyd sintering machine used at the Mines Experiment Station in 1953. Cakes of sinter may be seen at the extreme right.*

ONE OF THE MANY TACONITE *equipment flow sheets set up in the Mines Experiment Station's laboratory about 1930. The ball mill is in the center.*

A LATER FLOW SHEET *used about 1940 at the Mines Experiment Station to test east Mesabi magnetic taconite. Magnetic separators are at the extreme left.*

BALLING *the fine concentrate in the Mines Experiment Station's temperamental balling drum during the first continuous pelletizing demonstration given in April, 1943.*

SCREENING *the green balls as they emerged from the drum in the 1943 demonstration.*

BALLS *made in this furnace in 1943 were hard and porous but not uniform in size.*

THE RECTANGULAR *shaft-type pelletizing furnace used in the 1943 demonstration.*

By *1945, after Henry Wade devised a leveling bar for the balling drum, the Mines-Experiment Station got much better results.*

A TRAVELING GRATE PELLETIZING MACHINE, *designed for updraft burning, was used in a series of unsatisfactory experiments at the station about 1950.*

REPLICAS *of the "Lynn Pot" were cast from the first pig iron smelted from taconite pellets.*

FRANK E. VIGOR *of Armco sent a crew of blast furnace operators to the Mines Experiment Station to make the test in 1948. Courtesy Armco Steel Corporation.*

THE MINES EXPERIMENT STATION'S *blast furnace produced 150 tons of pig iron in 1948 in the first test of the smelting qualities of taconite pellets.*

ing the members of the legislature—particularly those on the appropriations committees—to act favorably on those requests. William K. Montague once introduced me as a speaker at a meeting in Hibbing by saying, "The geologists tell us that nature laid down the taconite here on the range a billion years ago. The next day Ed Davis started trying to sell it to the steel companies." I suppose I was a pest on the subject, but that was my job at the time.

The objective of the Mines Experiment Station's research on taconite had been to develop a process by which a superior product could be made. By 1944 we had demonstrated that this could be done, but we had not devised equipment for commercial operations. During the years we had been working on this process, certain far-reaching changes had taken place in the iron mining and steel industries. When we began the study in 1925, no one else was interested in Minnesota taconite, commercially or otherwise. The decision that faced us at that time was whether we should abandon taconite temporarily and wait for further depletion of the available natural direct-smelting ores of the Mesabi Range, or whether we should try to step up our laboratory investigations of taconite processing. Fortunately, the university allowed us to expand our studies. If we had waited for the steel or ore-producing companies to develop taconite processing, we would probably still be waiting.

As it turned out, a feasible laboratory process was on hand by the mid-1940s when the industry came face to face with the heavy drain on ore reserves caused by World War II. It was the Mesabi Range which enabled the United States to become the world's largest manufacturer of steel and made Minnesota the largest producer of iron ore in the nation. Because of the rapid depletion of the Mesabi natural ores, United States steel firms in the 1940s and 1950s were looking the world over for new ore reserves. They were successful beyond all expectations. Great new deposits of natural ores, capable of supplying the needs of the steel industry for many years, were discovered beyond the boundaries of the United States in Canada, South America, and elsewhere. Moreover, the long-discussed St. Lawrence Seaway was at last on the way to becoming a reality. Such a waterway could carry foreign ores to the steel plants along the Great Lakes, unloading at the very docks where Mesabi ores had always been delivered. If the vast foreign deposits and the St. Lawrence Seaway had become available earlier, or if the university had decided to drop the taconite study, I believe that taconite would probably still be just an interesting possibility. As it was, an alternative was available to steel firms seeking greater ore reserves. That alternative was taconite, and it offered the hope of a domestic rather than a foreign ore reserve. The pellets also gave promise of being a superior blast furnace product. Could they compete? In the 1940s the answer was by no means clear, but by that time the Mines Experiment Station no longer worked alone on taconite.

5 Taxes and Taconite in Minnesota

WHILE THE MINES EXPERIMENT STATION'S laboratory studies were progressing, other developments of importance to taconite were taking place. Late in 1938 the first of a series of events occurred that profoundly affected the whole taconite program. It all started very simply with a visit to the station by Oscar Lee, an engineer on the staff of Republic Steel Corporation and an alumnus of the University of Minnesota School of Mines. Oscar was studying a property which Republic had recently taken over near Mineville, New York. The mine there had been in operation for many years. The ore was being concentrated at a coarse size by dry magnetic separation, making a product of indifferent quality but coarse enough to require no agglomeration. Oscar's idea was to crush the ore finer, change to a wet process of concentration, and thus materially improve the quality.

He arranged for two carloads of ore to be shipped from New York to the university, and we made a series of fine grinding and wet magnetic concentration tests on it.[1] Very satisfactory results were secured, which were reported in May, 1939, to Donald B. Gillies, vice-president of Republic. As a result, the company decided to erect a pilot plant to try the process on a commercial scale. The necessary machinery was assembled and installed in an old mill building at Port Henry on the shore of Lake Champlain a few miles from Mineville. When the plant was ready to go into operation in the summer of 1940, experts and interested parties, including the author, were invited to be present at its opening. Everything went smoothly, and the plant produced concentrate which was low in phosphorus and silica and contained over 70 per cent iron.

After the ceremonies, Mr. Gillies asked me to stop in Cleveland on my way back to Minneapolis. He would call a meeting, he said, at which he would like me to describe the operation of the pilot plant for other officials of the company. Among those at the meeting to which Mr. Gillies took me after I reached Cleveland was Charles M. White, who was at that time vice-president in charge of operations for Republic. Mr. White seemed very pleased with the success of the wet concentration method in the new plant. When I was ready to leave, I remarked that he should go to Port Henry and see what a beautiful process fine grinding and wet magnetic concentration was. Then, I added, he should visit Minnesota, where we would show him a really big magnetite property.

Mr. White is very tall — well over six feet. After my remark, he looked down at me and, patting me on the shoulder, he told me to go back to my laboratory and have a good time with that taconite, but not to expect him to have any interest "in that God-damned hard stuff or anything else out there in Minnesota until you get over the idea of taxing everything to death."

Well, his statement startled me and gave me a lot of food for thought. Here we were working out a new process to make taconite more desirable, and an executive of one of the nation's most progressive steel firms wanted no part of it because of Minnesota's tax laws. When I inquired further, other steel men indicated a similar attitude, although perhaps not so positively expressed. Obviously, even if we at the Mines Experiment Station did develop an economical way of processing it, taconite was not going to be utilized until some changes were made in the mineral tax laws of the state.

At that time (1940) three different types of taxes were levied on iron ore in Minnesota: the ad valorem tax, the occupation tax, and the royalty tax. The combined total of these was large, but it was the first that was most objectionable to mine owners, who considered it unreasonable and unjust. It required that a mine property be taxed according to its valuation from the time of discovery until its ore was all gone. Rural property in Minnesota was then assessed at 33⅓ per cent of its full and true value, city real estate at 40 per cent, and mining property at 50 per cent. The valuation was set each year by a tax assessor, and then the mill rates of the various taxing districts in which the property was located were added together to form the total assessment.[2]

It had become the custom on the Minnesota iron ranges to extend the corporate limits of the various cities and villages to include adjacent mines and undeveloped mineral properties, and municipal limits had been drawn out until they covered practically the entire iron formation. Therefore, any mining property had levied against it the total of the mill rates established by the county, village, and school district in which it existed—like city property—plus the special taxes levied by the state on iron ore. To the mining firms, the most objectionable feature of the system was that a ton of ore in the ground was taxed in this way year after year until it was mined and shipped out of the state.

The effects of the tax situation went something like this: if a landowner on the Mesabi Range found ore on his property, the state, county, municipality, and school district would tax it at a high rate whether he mined it or not. Consequently, he was under great pressure to sell it to a company that could afford to pay the taxes on it year after year until the ore could be mined and shipped. The better the ore and the larger the deposit, the higher the valuation placed on the property by the assessor, and the sooner the owner must get it into production. This meant that the better ores from the best properties were mined first, and the poorer ones were left until the high-grade, high-tax ores were gone. As a result, little effort was made to mine the poorer ores until the rich ones were shipped, although mining the two would have been sound conservation practice worthy of encouragement by state and local tax laws.[3]

The ad valorem tax has been the very lifeblood of the range communities, and its revenues have for years provided them with many of the luxuries as well as the necessities of existence. The operation and financing of the communities has been planned around this tax, and attempts to change it have encountered violent resistance. As the ore in a municipality is mined out, however, taxes from this source decrease, and many of the range communities are now, or soon will be, faced with the necessity of reducing their standards of living, or, alternatively, taxing their own residents for the beautiful and expensive schools, auditoriums, and parks originally financed and for many years supported by the ad valorem tax on unmined iron ore. In 1940 steel men told us that they hesitated to locate plants or factories in these areas, because they feared the heavy tax burden would be shifted to them after the remaining high-grade iron ore was exhausted.

The steel and mining companies operating in Minnesota had seen much of the money they paid in taxes dissipated in local extravagance. They had also seen it breed the belief that the mining companies owed the people of the range communities a living—that they had brought the people into the area and were obligated to take care of them and their families as long as they remained there. You were either a friend or an enemy of the people, depending upon whether you were for or against the "steel trust" in any discussion.

When the Mesabi Range was opened in the 1890s, the mining companies imported large numbers of unskilled, immigrant laborers. At that time steam shovels on rails dug the ore and placed it in hopper-bottomed railroad cars pushed around by steam locomotives. It took a crew of a dozen men for each shovel, bringing coal and water, shifting tracks, and leveling and clearing the ground. Employment was largely unskilled, but plentiful, from May to December during the shipping season. For the rest of the year, mining operations ceased, and most of the men went into the woods to work for the lumber companies or to cut pulpwood.

By the 1940s the employment situation had changed radically. Electric shovels on caterpillar treads had largely taken the place of the steam shovels, and trucks had largely replaced railway cars in the pits. As a result, the number of employees needed in the mines had been greatly reduced. Younger men who could develop the skills required to become mechanics or truck drivers were given the better-paying, year-round jobs. During the winter when ore shipments ceased and navigation was closed on the lakes, the railroad and mining companies employed as many men as they could in the repair shops. But work was scarce, and large numbers of men remained unemployed. The timber was largely gone by that time, and even pulpwood supplied little employment. It was not a healthy situation, and the public relations programs of the mining companies in 1940 had not been developed to cope with it.

In part, too, the bad feeling dated from the early days on the range when the companies, controlled from the East, left local public relations in the hands of the mining captains. They were harsh, tough men both on and off the job; they made their own rules and enforced them with strong-arm methods. They were selected by the "steel trust" to open the mines and get out the ore, using large numbers of workers who were regarded as little more than work animals by their bosses. The captains were "company men," and the workers lived in "company towns," which were at first not very attractive places, poorly provided with such basic facilities as sewers and pure water. The workers' dislike or even hatred of the mining companies' local representatives was later transferred to the firms, and this feeling, coupled with high accident rates, irregular employment, and poor working conditions led to bitter but unsuccessful strikes in 1907 and 1916. Although union organization of the miners did not come about until the 1930s, the range communities from the beginning were sharply divided. Distrust between company men and other groups seems almost to have been fostered and promoted by both sides. In such an atmosphere, the children of the early immigrant workers grew up, and the heritage of bitterness persisted.[4]

Into this situation in 1913 stepped Victor L. Power, who in that year was elected mayor of Hibbing. He had grown up in Calumet, Michigan, and had moved to Minnesota as a young man to work as a mine blacksmith in the Hibbing-Chisholm district of the Mesabi. Later he attended Kent College of Law in Chicago and the University of Minnesota and passed the state bar examination in 1904. He practiced law in Chisholm and Hibbing and was mayor of Hibbing for nine terms (1913–22). On his death the *St. Paul Dispatch* of April 6, 1926, had this to say about him: "Mayor Power assumed office when Hibbing was little more than a mining camp. When he left the office of mayor, Hibbing was one of the most modern villages in the world. . . . Working on the theory that employes of the mines were entitled to all the civic benefits they could obtain, Mayor Power, with the aid of the village council, imposed heavy taxes on the mining companies, building costly schools, paving streets into the forest, lighting the streets for miles where there were few homes, providing the most adequate fire protection and building an expensive park system." [5]

With Power as its mayor, Hibbing advertised itself as "the richest little village in the world" and set the pattern that other communities on the range soon followed. With such large amounts of local tax money available there was bound to be greed and graft, and many were the stories of extravagance and waste — some true and some fiction — but believed or rejected depending upon whether you were a friend or foe of the "steel trust." [6] It was said that the Hibbing schools had cut-glass doorknobs, and the flagpoles were covered with gold leaf five times a year. It was also said that all members of the school board and the three shifts of janitors had Pierce-Arrows, which were serviced in the school garage, that players on the village baseball team were carried on the payroll as painters and to justify their salaries they painted the bandstand over and over again all summer long.

Many attempts were made by the mining companies and the Minnesota legislature to curb the extravagant spending of the range communities. The most effective effort resulted from a report prepared by a legislative Interim Commission on Iron Ore Taxation appointed in 1939. Senator Frederick J. Miller of Little Falls was chairman of the group until his death in 1940. Then Representative Rollin G. Johnson of Forest Lake took over, and Senator Karl G. Neumeier of Stillwater was elected vice-chairman. Professor Elting H. Comstock of the

University of Minnesota School of Mines was the commission's secretary. The group made a thorough study of conditions, held hearings in several range communities, and prepared a comprehensive report for the 1941 legislature. Its members were impressed by the necessity for putting a tighter limit on the expenditures of the range communities. They reported, for example, that Hibbing in 1938 had a total tax levy for village purposes of $1,370,401, or $87.47 per capita, while Mankato in southern Minnesota with about the same population had a levy of only $210,274 or $14.98 per capita. "In 1939 it cost Gilbert $113.78 per pupil for teachers," the commission stated, "and $99.05 per pupil for janitors, whereas at the same time it cost Waseca $50.35 per pupil for teachers and $9.18 per pupil for janitors." Gilbert had 763 pupils and Waseca 759.[7]

Perhaps the most important result of the commission's work was the liberal education received by its members from their contact with the mining industry and the people living on the ranges. Before the discussion of the commission's report, some legislators showed an appalling ignorance of iron mining activities. They knew there were great mines in the northern part of the state and that ore was taken from them. They knew the iron did not come from the ground in the form of nails, but they believed it was similar to the gold nuggets they saw being washed out of the sand in movies of the California gold rush — pure gold but not yet in the form of gold rings. Starting with that idea, a person could work up quite a sweat over how the mining companies were robbing the people of the state and how these firms could well afford to pay more taxes. "Why, they say that a ton of 60 per cent pure iron ore is worth only four dollars, but just a hundred pounds of nails cost me twice that much. Who gets all the gravy? It certainly isn't the people who work their fingers to the bone digging the stuff out for them." The fact was that the steel and mining companies had not taken the trouble to explain to the public the simple fundamentals of their operations. Even today, in all probability, a comparatively small percentage of the people living in the state understands the difference between iron and iron ore or between melting and smelting.

This was the situation when Mr. White of Republic Steel refused to have anything to do with Minnesota taconite. The whole atmosphere on the range was antagonistic to the steel industry. When a new mining property was to be opened, the first question usually asked by people living in the region was, "How much tax will it pay?" not "How large will its payroll be?" Industry was not much interested in moving into or expanding in a climate of this kind, and in the 1940s many steel companies started development projects in Labrador and South America rather than on the Mesabi.

At the university we realized that the application of the ad valorem tax to taconite would be disastrous. If only a minimum valuation and an average range community mill rate were applied to the estimated fifty billion tons of taconite on the Mesabi, the total tax might well be fifty to a hundred million dollars a year, and this would continue for years. At any conceivable mining rate, much of the taconite would remain in the ground for fifty to a hundred years, and the ad valorem tax would be collected year after year throughout that time. It seemed obvious to us that if taconite were to be utilized, some changes would be required in Minnesota's mineral tax laws, but we knew that violent opposition to any change in the ad valorem tax law, which was the chief culprit, was certainly to be expected in the range district.

Knowing the critical attitude of residents of the ranges toward the iron ore industry and the industry's equally critical attitude toward the state's mineral tax program, I consulted some members of the Junior Chamber of Commerce in the mining communities and also a group of university people in 1940–41 to see what, if anything, could be done to improve the outlook for taconite. Among the latter were: William T. Middlebrook, comptroller, and Professors Schwartz of geology, Frederic B. Garver of economics, and Horace E. Read of the law school. From these consultations the conclusion emerged that in order to revive interest in taconite, a law should be passed exempting it from ad valorem taxation. But any law passed by one legislature could be repealed by the next. It was difficult to believe that a hardheaded steel company executive like Mr. White would consider such a law anything but a "come on" to get taconite plants built, after which the tax base would be increased.

Since iron ore taxation is covered by a 1922 amendment to the Minnesota Constitution, we considered trying to obtain approval of another constitutional amendment that would exempt taconite from the ad valorem tax. The idea was abandoned, however, because it is very difficult to pass a constitutional amendment in Minnesota. It must be approved by a majority of the voters, and any voter who does not vote on the amendment is counted as voting "no." To obtain approval, it would be necessary to reach almost every voter and get him to understand the proposal and the difference between taconite and other iron ores. This seemed like an impossible task at the time. Certainly it would have taken years of education and thousands of dollars. After all, who was interested in taking taconite off the ad valorem tax rolls? The mining companies were not. They had plenty of ore outside the United States that was much better than taconite. The people of the range did not understand the problem and could be expected to oppose any proposal that even appeared to favor the mining companies. There were some very vocal critics of the "steel trust" in Hibbing and Chisholm, and the only friends taconite had in 1940 were the small group at the university, some members of the Junior Chamber of Commerce on the Mesabi Range, and a few scattered supporters of a low-grade ore program for the state.

The situation was very discouraging. Professor Garver thought an educational program was needed, beginning in the schools, but there wasn't time for that. The most feasible suggestion came from Professor Read. He said that if the people of the range communities asked to have a law passed removing taconite from the ad valorem tax rolls, such a move would be convincing to big business, since the residents of the area were really the only people seriously concerned. Although this was a constructive suggestion, it was a long shot both ways. Would the range people accept such an idea and sponsor the law? If they did, and a law were passed, would the steel industry accept it and invest its millions in Minnesota? We decided to make a try. The plan was to make this a range project with the university co-operating. No one expected the mining companies to take an active part, since at best mining people had no interest in taconite and at worst they were frankly scornful.

Early in July, 1940, I betook myself to the Mesabi to address the annual meeting of the Virginia Junior Chamber of Commerce. In the first of many speeches in 1940 and 1941, I presented charts and lantern slides to illustrate the fact that the reserves of high-grade ore on the Mesabi were declining, that an enormous tonnage of taconite was available, and that increased employment would result from any large-scale use of this resource. I told the audience that the future of the Mesabi Range "lies in your ability to use low-grade materials," but that this would never come about without a change in the ad valorem tax law. The talk was fully reported in the local papers, and I got back to Minneapolis all in one piece.[8]

It so happened that Wayne Apuli was the Mines Experiment Station's chief chemist at the time, and one day in the fall of 1940 he brought his father, a Mountain Iron merchant, to the laboratory to show him around. Gustave A. Apuli was a shrewd, intelligent individual, well acquainted with range politics and problems. We talked about the need to conserve high-grade ores and about taconite and the ad valorem tax, and he said that he would see what he could do to help. In a short time word came from him that he would like to bring to the station some interested visitors from the Virginia-Mountain Iron district. This was arranged, and Mr. Apuli soon arrived with Grove Wills, owner of the *Eveleth Clarion,* William A. Fisher, publisher of *Range Facts* in Virginia, and Henry G. Tiedeman, superintendent of schools at Mountain Iron. We showed them the taconite processing flow sheet we had worked out and gave them a lot of statistical information on the depletion of the area's rich ores and what this would mean in the loss of jobs, as well as taxes, unless mining companies could be encouraged to utilize low-grade materials and taconite.

As a result of their visit, this group of influential men invited others to attend a meeting on December 2, 1940, at the old Fay Hotel in Virginia, for the purpose of hearing what I had to say about the future of iron mining on the Mesabi. Most of the state senators and representatives from the area were present, as well as a number of younger men who were later to play an active role in state and national politics.

According to some old clippings in my files, when they gave me the floor, I told them about the tremendous reserves of taconite in their area and what these could mean to the range in terms of employment. Using charts and lantern slides, I described processing methods and traced the development of Western low-grade copper ores as an example of what might be done with taconite, giving figures on the cost of constructing and operating such plants. At that time it seemed that a million-ton taconite plant would cost ten to twenty million dollars, and no firm would invest such a sum until a more favorable tax program for taconite was enacted. I said that the range communities were receiving no taxes from taconite, and they never would, and no payrolls either, unless something was done to show at least the good faith of the area. No action was taken at the meeting, but the men listened to my rantings attentively.[9]

Later that evening in the bar of the Fay Hotel I had discussions with two men who were to be candidates for election to the Minnesota legislature — John A. Blatnik of Chisholm, now a member of Congress, and Thomas B. Vukelich of Gilbert, now a state senator. Grove Wills, Howard W. Siegel, an Eveleth lawyer, Gilbert P. Finnegan, postmaster of Eveleth, and others were also present. It was decided that if the university people would submit a proposal to take taconite off the ad valorem tax rolls, these men would support it. In the meantime, they would spread around the idea of passing a tax law that would make taconite more attractive to industry.

Everyone understood that the ideal procedure would have been to work for a constitutional amendment, but all agreed that passage was almost impossible. The next best thing was to get the people on the range — the most tax-conscious citizens in the state — to sponsor a law that would remove taconite from the ad valorem tax rolls. However, if this proposal did not receive virtually unanimous support, that is, if there were any strong organized opposition, we might as well drop the whole matter. It was pointed out that we were asking quite a lot of the range people. We were asking them to sponsor a law prohibiting themselves from taxing taconite as they had the rich ores. But there was no alternative. Something of this kind was needed as convincing evidence that Minnesota was really in earnest about taconite and would not try to change the law again as soon as big investments had been made. Any indication of the attitude "Now we got 'em, let's soak 'em" would be fatal.

It was a good meeting that night at the Fay, and for the first time we had a definite program. We at the university were to prepare a new tax bill that range residents could support in a manner that would be convincing to the steel people of the East. Then some of the legislators from the area would introduce it and attempt to get it passed and signed by the governor. That was quite an order in itself, since at the time the state government was entirely Republican and the range was strongly Democratic.

Back at the university a question came up about the constitutionality of exempting unmined taconite from ad valorem taxation. After some discussion, it was decided that if any lawyer could draw up a constitutional measure it would be Bill Montague, a former range resident who had served for a time (1929–31) in the state attorney general's office. He later moved to Duluth, became secretary of the Lake Superior Industrial Bureau, and was therefore familiar with the problems of the mining industry as well as the requirements of the state constitution. Mr. Montague was approached in December, 1940, and agreed to draw up a bill that would be constitutional. He did so, eventually sending to the university a draft which we modified and altered in its details but not in its principle.

We had a lot of trouble preparing a definition of taconite. At first, the geologists made it very technical and rejected all attempts to simplify it. As a compromise, two definitions were written; one was geologically acceptable and the other was intelligible to laymen, and both were placed in the bill.[10] The tax rate figures which were to apply to taconite also were discussed in detail, and finally a sliding scale was accepted of five to six cents per ton of shipping product, depending upon grade. But the basic idea supplied by Bill Montague, of including a small ad valorem tax to satisfy the constitutional requirements and then applying a heavier tax on the concentrate shipped, was retained just as he prepared it.

Copies of the bill were submitted to all range legislators at the 1941 session, and after careful consideration, Representatives Vukelich and J. William Huhtala agreed to sponsor it in the House. Mr. Blatnik was to shepherd it through the Senate.

On the range, the Junior Chamber of Commerce members took up the program, supported by such newspapermen as Laurence A. Rossman, Sr., of the *Grand Rapids Herald Review*, Burt D. Pearson of the *Virginia Daily Enterprise*, and George A. Perham, editor of *Range Facts*. With information supplied by those of us at the university, these men were apparently successful in creating favorable sentiment on the range for the proposed legislation—at least no real opposition developed. But some people certainly made fun of the whole idea. They called Blatnik and Vukelich "the taconite twins." At one tax committee hearing on the bill, after listening to me orate on the declining mining industry and the importance of taconite, one member of the committee was overheard telling another, "I think the old boy actually believes all that stuff."

The bills were introduced by Liberal members of both the House and Senate, but we knew it would not be possible to pass them without Conservative support. (The Minnesota legislature is organized on a nonpartisan basis into Conservative and Liberal caucuses. Broadly, the Conservative members tend to reflect the views of the Republican party, the Liberal those of the Democratic Farmer Labor party.) The tax committees of both chambers were heavily Conservative, and a Republican, Harold E. Stassen, was governor. As it happened, an opportunity arose for some of us at the Mines Experiment Station to be of service to Rollin Johnson and Karl Neumeier, the chairmen of the House and Senate tax committees. These men, who had served on the Interim Committee on Iron Ore Taxation, had introduced bills to place a per capita limit on local taxation. A lot of opposition developed in the House, and Johnson, who was not a mining man, was having trouble answering the arguments. John Craig and I sat down with him and gave him a lot of statistical information on iron mining and answered many of his questions. Then we took the opportunity to explain to him the importance of the taconite tax law that would be introduced by Representatives Vukelich and Huhtala and that would come before his committee. After much debate, the legislature passed Johnson's bill limiting local taxes to a maximum on a per capita basis.[11] He was appreciative of the help we had given him, and in return agreed to press his committee to approve the taconite tax bill.

In the Senate, Karl Neumeier and his committee were engaged in a big public wrangle with Governor Harold Stassen. The governor had made a speech on the radio announcing his intention to present a new bill that would rebate part of the occupation tax to mining companies operating high-cost properties. The committee had been studying a similar proposal, but Governor Stassen made his announcement without consulting it. The tax committee members were determined not to let the young governor get credit for this idea, and they wanted to present a substitute measure of their own.

We met with Senator Neumeier and agreed that we would help him work out a bill the committee could sponsor which would differ from the governor's but would accomplish the desired purpose of providing some tax relief to the low-grade ore mining industry. In return, the committee was to give favorable consideration to our taconite tax bill. Some of us, especially John Craig, spent many hours working out various formulas and showing statistically how, in past years, these would have rebated tax money to the small operators, while the governor's program would have rebated tax money principally to the big companies. Senator Neumeier selected one of our formulas and wrote a tax bill around it that was passed in place of the governor's bill. It became known as the labor credit law and was still in force in 1964, rebating taxes on the basis of employment and payrolls.[12] We felt that this law would also help future taconite producers, who would be big employers having very high payrolls per ton of product.

Governor Stassen heard about the work we were doing for Senator Neumeier's committee, and he called me in to ask why we were opposing his program. When he heard about the deal we had made to get the Republicans to pass a taconite tax bill sponsored by the Democrats, he simply smiled and said, "Perhaps I will not sign it."

Rollin Johnson and Karl Neumeier took the taconite bill through their committees all right, but when it reached the Senate floor a serious hitch developed. Under Minnesota law, the House must originate and approve all revenue-raising measures before the Senate can act on them. John Blatnik was a new senator that year, and on the evening of April 17 he received word that the House had passed the taconite tax bill. He then asked to have the Senate bill called up, and it was quickly approved without a dissenting vote. But a little later it developed that the House had not yet passed the measure. The records show that it did so the following day.[13]

Whether John had been informed of the House action by a friendly senator making an error on purpose or by honest accident may never be known, but the effect was the same. The Senate bill had not been legally passed. There was not sufficient time remaining in the session to return the bill to committee and bring it up again for another vote. Only one possible way of saving it remained. Karl Neumeier was a very influential Conservative in a strongly Conservative Senate, and by exercising the privilege and prestige of his position as chairman of the tax committee, he could ask the Senate to repass the bill as it was sent over by the House. When the situation was explained to him, however, he at first refused. He and Blatnik had been on opposite sides of the debate on the per capita limitation bill, and many hot and uncomplimentary words had passed between them. Why should he rescue Blatnik's bill for him? After we reminded him of the agreement he had made with us, Neumeier reluctantly consented. In the evening session on April 18, the Senate suspended the rules requiring that a bill be read three times and passed the House bill unanimously. This time everything was legal. To our relief Governor Stassen promptly signed the measure, and by April 23 Minnesota had its first taconite tax law.[14]

Another bill affecting taconite development was proposed by the university people at this same session of the legislature. When the territorial government of Minnesota was established in 1849, two sections of land in each township were reserved for educational purposes. (More lands were later acquired by other means, and the income from them goes into a state trust fund, which is used partly to support the public schools and partly for the university.)[15] In this way, the university became the owner of considerable land on the Mesabi Range, some of which contained iron ore but most of which was taconite. The state had a system of leasing mineral land to mining companies which had been designed for direct-shipping ores and was entirely unrealistic when applied to taconite or other low-grade iron ores. We who were working on taconite wanted the provisions changed so that industry might be interested in leasing the university's taconite lands.

The Minnesota Division of Lands and Minerals, under the direction of Ray D. Nolan, agreed to prepare a bill to accomplish this. After much study the division presented, and the 1941 legislature passed, a bill that revised the entire mineral-leasing program of the state. With only minor changes, it is the law under which mineral lands were still leased in 1964. In general, the law provided that the successful bidder must pay the state an annual ground rental charge plus a royalty on the ore produced, the minimum acceptable bid depending upon the nature of the ore and the type of mining operation required. The royalty was to be based upon the mineral content of the product shipped and sold—not entirely upon the quality of the material mined. Thus taconite mined from state lands pays a royalty on each ton of concentrate produced rather than on each ton of crude rock mined.[16]

Certainly many members of the legislature in 1941 thought these laws were of little importance. I am afraid most people on the range regarded them as visionary and felt that no mining company would ever use taconite anyway. They dubbed the ideas that Minnesota's high-grade, "easy ores" were fast disappearing and it would be necessary to use the low-grade materials as "just mining company propaganda" invented to scare them. For years they had heard talk that the big ore firms were hiding from the tax authorities vast deposits of high-grade ore which they would "discover" as needed.

The 1941 taconite tax law had no backing from any mining company, which is probably the reason the iron range communities and their legislative representatives were willing to approve and sponsor it. Its enactment put Minnesota on public record as inviting the utilization of taconite. It also allayed the fears of existing mine operators that the taconite surrounding their mines might be taxed in the ground like the high-grade ores. Whether or not these two measures would offer enough incentive to get a taconite industry started remained to be seen. At any rate, I took great pleasure in sending copies of the new laws to Mr. White of Republic Steel.

6 | Taconite Comes to Life Again

In the years immediately following the passage of the 1941 laws, the taconite giant stirred and shook himself. The support of the 1941 measures, especially the tax law, by the people of the range communities was apparently convincing proof of their sincerity, for the legislature had scarcely adjourned when several large mining companies began to take a more active interest in Minnesota taconite.

Erie Mining Company, an organization formed in 1940 and managed by Pickands Mather for four steel firms, was the first to act.[1] Pickands Mather, whose business is to provide its affiliated steel companies with adequate ore reserves, had long taken a passive interest in Mesabi taconite. Encouraged by president Elton Hoyt II of Cleveland, John C. Metcalf of the firm's Duluth office had been a frequent visitor at the experiment station, where he watched our taconite processing equipment in operation.

In 1941 Erie began to take leases on state-owned taconite lands on the east Mesabi,[2] and in 1942 it set up a new laboratory at Hibbing for testing samples. Large sections of the land leased by Erie were University of Minnesota trust fund lands, and because of the active part we at the university had played in the passage of the taconite laws, it was soon whispered that some sort of deal had been made and that money had changed hands. The writer of these pages was in a position to know everything that occurred and can state positively that there was absolutely no collusion or pay-off of any kind.

The Mines Experiment Station staff was certainly not in a position to gain anything. As a matter of fact, Erie told us little about what was being done in its laboratory and later locked the doors of its pilot plant against us. We learned, however, that the firm was trying various methods to produce high-grade concentrate from its newly acquired east Mesabi taconite properties. Its Hibbing laboratory was under the direction of Fred D. DeVaney, a 1923 graduate of the university. He had formerly worked for the United States Bureau of Mines, where he became an expert on flotation. DeVaney and his staff thoroughly investigated the flotation of Erie's taconite and developed a successful concentration method. Later, however, this line of research was set aside because it appeared that the cost would be high, and the firm worked with the Mines Experiment Station to install equipment and test the magnetic separation and pelletizing process we were developing.[3]

Reserve Mining Company at first moved somewhat more slowly than Erie. It will be remembered that Reserve had been incorporated by Oglebay Norton in 1939 and had acquired the holdings of the old Mesabi Iron Company, including the plant at Babbitt.[4] Reserve's first president was Crispin Oglebay. He was also chief executive of Oglebay Norton and a grandson of that firm's founder. A man of wide vision, his imagination was fired by the possibility of developing from east Mesabi taconite an even greater supply of ore than had been produced from the rich mines of the Mesabi Range. Aware of the fast-disappearing reserves of high-grade ore in the United States, he was concerned about finding other domestic sources so that the country need not depend on foreign ores for such a basic commodity as iron.

In organizing Reserve, Oglebay had a kindred spirit in Charles R. Hook, chairman of the board of Armco Steel Corporation. When Reserve was formed in 1939, these men were not thinking of the immediate future. They called the new company "Reserve" because they thought of taconite as a long-time future resource that might not be needed for twenty years, but would be there waiting when the right time came. Neither Oglebay nor Hook anticipated any immediate use for taconite, and at the time they had no men in their organizations who were experienced in iron ore concentration.

But events moved faster than the steel men expected. The United States entered World War II in 1941, and heavy demands were placed on the remaining high-grade Mesabi ores. After the Minnesota legislature acted to encourage taconite production, Henry K. Martin was added to Oglebay Norton's staff in the spring of 1942. He had been graduated from the University of Minnesota School of Mines and had been employed by the Colorado Iron Works. He was acquainted with the fine grinding and classification of ores, especially in large Western copper plants. Martin's appointment by Oglebay Norton was the first sign we at the Mines Experiment Station had that the firm had any immediate interest in taconite. After he joined the company, Martin spent a great deal of time at the station working with us on the development of the taconite flow sheet. His valuable suggestions contributed greatly to the success of the process.

There was a brief flurry of interest in the abandoned Babbitt plant during World War II. In 1942 the War Production Board, which was organized to see that raw materials required for the war effort were available as needed, became concerned about the nation's supply of iron ore. Two of the board's members at the time were R. C. Allen, vice-president of Oglebay Norton, and Charles K. Leith, one of the geologists who had studied the Mesabi Range in the early 1900s and written for the United States Geological Survey two monographs that are still the basic references on the geology of the district. These men asked what would be needed to put the old Mesabi Iron Company's plant into operation again. By the time the data had been assembled, however, the composition of the board had changed, and its members decided that, since the plant could not be put into production for about a year, they were not interested. The decision was a great disappointment to those of us in Minnesota who hoped to rejuvenate taconite processing. As things worked out, however, perhaps it was just as well the board voted as it did, because some technical problems regarding the process had not yet been completely solved.[5]

Other companies also began studying iron ore concentration methods in the early 1940s. Bethlehem Steel became interested in pelletizing in 1943. A year later the Oliver Mining Division of United States Steel established a large experimental laboratory in West Duluth to study taconite and other low-grade ores.[6]

In the late 1940s and early 1950s co-operation among the various firms was limited. Very little information was exchanged, some groups being more secretive than others. Erie and Oliver closed their doors tightly, and Bethlehem opened its with some reluctance. As a result, progress was delayed and there was much duplication of effort and expense. Erie and Reserve, whose properties actually adjoined near the old Spring Mine east of Mesaba, and who had similar problems of mining, concentration, and agglomeration, operated for a time with practically no technical contact between their engineers. Fortunately, this unhealthy situation lasted only a few years.

During this period, experimental work on taconite increased as three of the four companies and the Mines Experiment Station attempted to perfect methods of processing the ore. In the 1940s it was by no means certain that the flow sheets which worked in the laboratory could be adapted to commercial production. The equipment needed to make a few pounds of concentrate is quite different from that required to produce successfully millions of tons in a large plant. Furthermore, the increased size of a commercial operation creates totally new and often unanticipated problems. For example, where was the enormous quantity of water required by the process to be secured? How were the millions of tons of tailings to be disposed of? There was one other big question concerning taconite — a question that was not to be answered during the decade under consideration here. Even if the laboratory process turned out to be workable on a commercial scale, would it be economically feasible? Could a competitive taconite product be manufactured that steel producers would buy?

In the early 1940s the record-setting shipments of ore from

the Mesabi to satisfy the demands of the war and the taconite activities of Erie and Reserve drew the attention of the press. Trade journals, newspapers, and national magazines began to pose questions about "The Coming Crisis in Iron," as the *Saturday Evening Post* called it, the "Iron Ore Dilemma," in *Fortune's* words, and "Can Research Save Mesabi?" as *Business Week* asked.[7]

Many of the magazine writers raised questions about the costs of taconite processing. The answers we at the Mines Experiment Station could give in the mid-forties were not reassuring. In every estimate we could make there were two items that appeared to be excessive: the cost of electric power, and the cost of rail transportation from the Mesabi Range to Lake Superior. We attempted to make our estimates as realistic as possible, but about the best we could do was to show that a ton of taconite pellets delivered to ports on the lower Great Lakes would cost about two dollars more than the computed Lake Erie price. We were convinced that the improved smelting quality of the pellets would make up for a large part of the excess cost, but this was only an opinion and could not be reflected in our figures. So we took steps to see if the fixed costs could be reduced.

In casting about for some cheaper means of transportation, we learned that fine ore was being pumped considerable distances at some Western plants, and in 1942 Dr. Lorenz G. Straub of the university's department of hydraulics traveled to Arizona to investigate. At that time the Minnesota rail freight rate on taconite amounted to eighty-one cents per ton. Dr. Straub reported that finely crushed taconite concentrate in water could feasibly be pumped as a thin mud from Babbitt to Lake Superior for an estimated fifty cents per ton. In the West he noted that finely crushed tailings from copper concentrators were being pumped a distance of five miles across rough country. He estimated that, at worst, a series of twenty similar installations would be required to reach the lake from Babbitt. Our idea at the time was to crush, grind, and concentrate the ore at Babbitt, and then pump the concentrate to the lake, where it would be dewatered, agglomerated, and shipped. In this way we hoped to save thirty cents per ton on concentrate shipments to Lake

Superior and also avoid shipping to Babbitt the fuel required for agglomerating.[8]

The pumping idea seemed fine until someone asked what would happen if there were a power failure in winter. Obviously the line would have to be drained quickly before it froze or the solids settled in the pipe. A drainage outlet would have to be provided at every low point in the line, and a collecting basin and reclaiming system would be needed at every discharge point. In the West, where worthless tailings were being pumped, automatic dump valves were located at low points, and these opened and drained the line whenever the power went off. Taconite concentrate was too valuable to be spilled out on the ground and left there every time anything went wrong. We concluded that pumping was impractical in northern Minnesota's climate.

In 1944 when we approached officials of the Great Northern Railway, one of the large ore carriers on the range, to discuss a possible reduction in the freight rate on highly processed ore such as taconite, we received no encouragement that anything could be done. We were unsuccessful in getting the railroad executives to understand that if they would not reduce their rates, some competing form of transportation would be provided. A modern mining project of the scale required for taconite processing would be financially as strong as the range railroads and would be perfectly capable of building and operating its own transportation system. This had happened at large copper properties in the West and would certainly occur in Minnesota unless the existing roads made material concessions in transportation costs. In the 1940s rates of from eighty to ninety cents per ton were repeatedly sustained by the Interstate Commerce Commission regardless of the grade of product shipped. This meant that low-grade ores paid a higher rate per unit of iron than high-grade ores.

Fortunately we were more successful when we tackled the cost of electric power. At that time (1944) the rate for large consumers on the range was high, averaging over one and a half cents per kilowatt, perhaps because the demand varied widely between winter and summer. We arranged a conference with Richard F. Pulver of the Minnesota Power and Light Company, which furnished most of the power for the mining industry of the state. We discussed the rate problem for taconite processing with him in great detail; he was both interested and co-operative. As a result, the company established a special "taconite rate" of just under one cent per kilowatt. This allowed us to reduce our overall cost estimates for taconite processing by about fifty cents per ton of pellets. Although we had to continue to use the existing rail freight figures, the lower power rate brought us nearer the break-even point. We were sure the superior quality of the pellets would make up the difference by lowering smelting costs, and we thought we were getting a little closer to resolving the "Iron Ore Dilemma" *Fortune* had described.[9]

The Mines Experiment Station's work on grinding, magnetic separation, and agglomeration, which has been described in Chapter 4, was progressing for the most part along successful lines. Pelletizing, however, continued to give us trouble, and it was far from ready for commercial use. Not only did it require a large amount of fuel,[10] but it still presented a number of unsolved technical problems. In an effort to achieve both fuel economy and a better method of firing the pellets, we stepped up our investigations at the station in 1944.

We had learned that firing the green balls was fundamentally a very simple operation. They could be fired in many ways, but basically there were two general methods: applying the heat from below in updraft firing, or applying it from above in downdraft firing. We had first used updraft firing in a vertical shaft-type furnace. Such furnaces are efficient, simple to construct, and easy to maintain and repair. With combustion chambers on the sides, we had no difficulty in heating a column of pellets four feet thick.[11]

In this furnace, however, the pellets had a tendency to soften and fuse (or sinter) together if they were even slightly overheated. The least error in temperature control, or local overheating due to segregation or bad distribution, would cause fusion to start. Once begun, it seemed to get progressively worse until large masses of clinker (sinter) were formed in the furnace.

Some method of discharging them had to be provided, and this was not simple to do.

In an effort to solve the clinker and fuel problems, we constructed a new furnace at the station in 1946. It had a round, brick shaft three feet in diameter and seven feet high, but no combustion chambers. We devised for it an oscillating grate discharge mechanism which we hoped would take care of the clinkers. This kind of furnace had certain theoretical advantages as well as disadvantages over our earlier rectangular shaft type with combustion chambers. Commercially it would be less expensive to build, but some features of its design were highly experimental.

After many trials we learned that the round furnace would operate with air blown up from the bottom into which a small amount of natural gas was introduced as fuel. Eventually, however, we found that we got better results if a small amount of coal were mixed with the taconite concentrate before the balls were formed. Then we learned that if sufficient coal were added, no natural gas or other fuels were needed. The furnace made satisfactory pellets, but we still had trouble with large clusters and clinkers that could not be discharged and sometimes had to be broken with a crowbar. The oscillating grate mechanism worked effectively in breaking up some of the clinkers, but we still did not have a good way of discharging the really large masses that became solidly fused together. This turned out to be one of the biggest stumbling blocks in the entire process, and we were still working on it five years later.

Because of the difficulty of discharging the shaft furnace, we also investigated the possibility of burning pellets on a traveling grate. In a demonstration at the station in 1945,[12] we laid evenly a six-inch-thick bed of damp taconite balls mixed with about 7 per cent by weight of finely crushed anthracite coal on the moving grates of a small Dwight and Lloyd sintering machine. The grates on this machine are arranged to operate as a flat, endless belt which passes over pulleys or curved slides at each end. A fan draws air down through the bed of balls. We ignited the bed on the upper surface by a gas flame and burned downward, adjusting the grates' speed of travel to discharge the burned pellets or fused clinkers over the end of the machine as soon as the burning zone reached the grates.

This machine completely solved the problem of clinker discharge that had been so troublesome in the shaft furnace, but otherwise results were disappointing. After the bed of balls was ignited on the top, the downdraft evaporated the water in the damp concentrate near the top of the bed; the water then condensed on the cold balls near the bottom, causing them to soften and mash down into a tight charge through which air could pass only with difficulty. Thus the depth of the pellet bed that could be burned was very limited, and we became convinced that good pellets could not be made with such sintering apparatus involving downdraft firing.[13]

Early in 1950 we turned our attention to updraft burning on a traveling grate machine. Such apparatus would have advantages over a shaft furnace, chiefly in simplifying the feeding of the green balls and in discharging the fired pellets; it would also have disadvantages, we thought, in that its heat economy would not bc as good, and more maintenance would probably be required than for a shaft furnace. It remained to be seen whether the advantages outweighed the disadvantages.

We set up a research program, supported by both Erie and Reserve, to find out. This was the first break-through in the policy of secrecy that had prevailed between the two companies. Reserve provided a traveling grate machine, which we installed in 1951, and Erie furnished several carloads of taconite concentrate. Other expenses were divided three ways among the two firms and the university. We made a long series of tests, the results of which were reported to Erie and Reserve on July 27, 1951, March 31, 1952, and September 11, 1953.[14]

The machine had a capacity of about two tons an hour, and when conditions were right, it made good pellets. No attempt was made to recover the heat in the hot pellets, and therefore the fuel consumption was high, even higher than in the shaft furnace. The fuel used for ignition was natural gas, but a small amount of finely crushed coal was added to the concentrate before it was made into balls. There was a considerable lag between the time the coal was metered into the concentrate and the time the balls of concentrate reached the pelletizing machine. This delay made it impossible for the operators to quickly increase or decrease the firing temperature of the pellets as the operation might require.

To solve this problem, Wayne Apuli developed a method of rolling the fine coal on the surface of the balls as they left the balling drum. This was a real step forward for two reasons: it gave the pelletizing machine operators immediate and simple control over the fuel; it developed the heat on the surface of the balls where it was needed and prevented the pellets from overheating near the center. (We later learned that while we were working on the coal coating of the balls, a California firm had a detective at the station to keep it informed of what we were doing. When Reserve started to use coal-coated balls, the California outfit tried to claim damages.)

Even with this improvement, the tests were not completely successful. Although updraft burning on a traveling grate solved some of our problems, it created a few new ones. On the plus side — the furnace operators now had positive control of the fuel supply; a bed of pellets from eighteen to twenty-four inches thick could be burned updraft, compared to a bed only five or six inches thick in downdraft firing; and the pellets discharged easily whether they were clinkered or not. On the minus side — fuel consumption was higher than in our shaft furnace, and if the machine were to burn updraft, the bed had to be ignited on the bottom, where it rested on the traveling grates. Henry Wade worked out a way of doing this, but the design engineers of the two firms were never enthusiastic about it. Moreover, while the bed of pellets burned through completely, the sides tended to burn too fast because of excessive air, while the center burned slowly because of a shortage of air. (This is known as "wall effect," and it still makes trouble in commercial plants.)

As a result of this co-operative research program, Erie's engineers decided to continue with the shaft-type furnace. Reserve's engineers were impressed by the advantages of the traveling grate machine, and they continued to investigate this firing method. As it turned out, both methods eventually came into general commercial use, but much developmental work was required before this occurred.

On July 1, 1948, Erie began the operation of a pilot plant near its taconite properties at Aurora to test the taconite processing and pelletizing methods developed at the Mines Experiment Station and in the company's own laboratory. This plant was the first installation with commercial-size equipment to be constructed since the Mesabi Iron Company's ill-fated venture at Babbitt in the 1920s. Although it had a rated capacity of 200,000 tons of concentrate per year—larger than the old Babbitt plant—it was built for experimental operation only.[15]

In the Erie plant, the taconite from the mine was first broken in gyratory and cone crushers to particles of about five-eighths of an inch, and then fed to a twelve-foot rod mill. From this point in the flow sheet there was considerable flexibility, employing wet cobbers, magnetic separators of various kinds, and ball mills. The final concentrate—most of which was finer than 200 mesh—was filtered, and then conveyed to a balling drum eight feet in diameter and twenty-four feet long. The balls of concentrate passed over a section of the drum equipped to screen out the fine material and the undersized balls, which were then returned to the feed end of the drum for rerolling. Erie experimented with several types of linings in the drum to keep the concentrate from sticking, but later installed a leveling mechanism similar to the bar that Henry Wade had devised at the Mines Experiment Station. Normally, goods balls were made in the Erie plant.

The plant was equipped with a round pelletizing furnace eleven feet in diameter and sixty feet high—the first and only round commercial-size pelletizing furnace to be constructed. It had two combustion chambers and was designed to recuperate the heat in the pellets before they were discharged. Air blown into the bottom of the shaft passed up through the pellets cooling them and becoming preheated. It was then drawn into the combustion chambers following the design proposed by Charles Firth in his 1944 paper on pelletizing.[16]

As was to be expected in a new operation, various difficulties developed. These were not so serious in the crushing and concentrating stages (which eventually followed about the same steps used at the university), but in the pelletizing furnace the same old problem turned up with the discharge equipment at the bottom of the shaft. The pellets stuck together slightly, making some type of mechanical breaker essential. This had been omitted when the furnace was built, probably because it is difficult to design breakers for a large round furnace. The result was irregular operation; some pellets were too soft because of underheating, and some fused because of overheating. Occasionally the partially fused charge hung in the shaft and dropped suddenly, endangering the whole furnace structure as well as the operators. Eventually Erie abandoned the round furnace and constructed a rectangular one similar to the small model designed earlier by Firth, but with clinker breakers in the bottom.

Shortly before the Erie pilot plant went into production, Oglebay Norton, as operating agents for Reserve, organized a firm called Iron Ore Pelletizing Enterprise. The new company assumed the task of developing for Reserve commercial equipment for pelletizing the firm's fine magnetic concentrate. An experimental plant was constructed for this purpose in 1949 on the grounds of the Armco Steel Corporation at Ashland, Kentucky. Kenneth M. Haley, an Armco metallurgist, was in charge of the test plant, assisted by Donald E. Cooksey and James A. Reynolds.[17]

Since no taconite concentrate was available at Ashland for the operation of the test plant, magnetic concentrate was purchased from the Jones and Laughlin Steel Corporation, which had a taconite-type plant operating at the Benson Mine in New York. This magnetite concentrate was comparatively coarse, however, and Enterprise found it necessary to install fine-grinding equipment in order to simulate the concentrate that could be made from the lands Reserve controlled on the Mesabi.

A shaft-type pelletizing furnace, four feet wide by twelve feet high, was designed for Enterprise by Arthur G. McKee and Company of Cleveland. The furnace was constructed without combustion chambers, following the latest Mines Experiment

Station design, but we were again unsuccessful—as we had been with Erie—in persuading the engineers to design a furnace with adequate breakers in the bottom of the shaft. Moreover, the temperature in the furnace was controlled by adding natural gas with the air entering the bottom of the shaft. This was rather tricky; the amount of gas had to be proportioned exactly to the amount of air or an explosive mixture could result. On one occasion when the gas-proportioning valve did not function properly, an explosion did occur. It blew out the side of a fan, but fortunately did no other damage.

The Ashland test plant had a nominal capacity of about twelve tons of pellets per hour, but it never operated steadily at this rate. At first the pellets it produced were not especially good in quality. As they had in Erie's test plant, difficulties developed in adapting the university's laboratory process to commercial operation. Most of them grew out of the fact that no one as yet fully appreciated the absolute necessity of making strong, clean balls of uniform size, and of distributing them evenly and gently over the surface of the furnace charge. But changes in operating conditions and in the design of some of the equipment, especially in the balling drum and in the discharge mechanism at the bottom of the furnace shaft, improved results.

While the Ashland pelletizing test plant could hardly be considered a highly successful operating unit, it did, as planned, provide information of value. For one thing, it demonstrated that under proper conditions good pellets could be produced in a furnace without combustion chambers. Temperature control was difficult, however, and mixing natural gas with the air entering the shaft could not be considered safe operating practice in a commercial plant. No furnace of this type has since been constructed.

The Ashland plant also demonstrated the importance of maintaining exactly correct conditions on the inner surface of the balling drum. At first, the plant had a seven-foot drum lined with rubber. It proved unsatisfactory. Later the rubber was removed, and a reciprocating leveling bar was installed similar to the one Henry Wade had devised at the university. Ken Haley then put a variable speed drive on the cutter bar and redesigned its teeth. This worked wonders, for it gave accurate and immediate control over the nature of the surface upon which the balls of concentrate rolled. Haley's experiments indicated that the surface must not be too smooth or too rough, and a surface that was right for one type of ore might not work on another having a different moisture content or particle-size distribution.

Still another development at Ashland which has radically affected the operation and design of taconite plants was the discovery that fine damp concentrate could be successfully stored in large conical bins as it came from the dewatering filters and fed out as desired with a standard rotating table feeder. In the days of the old Babbitt plant, we believed that the filtered concentrate would pack down in a bin and could not be made to move through a feeder. Without such a bin, proper control of the feed to the agglomerating equipment is practically impossible. In modern pelletizing plants great effort is made to feed the filtered product to the balling drum at exactly the required rate and moisture content, for a change of a few tenths of one per cent in the moisture seriously affects balling characteristics and makes necessary drastic readjustments in subsequent operations.

During the three years the Ashland pilot plant operated, it produced 65,000 tons of pellets of varying quality which, mixed with other ores, were smelted in one of Armco's big blast furnaces. The amount available, however, at any one time was not sufficient to demonstrate the real value of the pellets.

While the Erie and Enterprise preliminary plants were struggling with balling and pelletizing problems, Percy L. Steffensen was at work in Bethlehem's Lebanon plant. For many years that company had been mining a magnetic ore somewhat similar to taconite at Cornwall, Pennsylvania, and shipping it to Lebanon for concentration. The coarse crystalline ore required crushing to only about 10 mesh, and the concentrate was being sintered. Mr. Steffensen had been at the Mines Experiment Station in 1943 when pelletizing was first demonstrated to the industry. An able and clever engineer, he at once recognized the importance of the university's process and also the difficulties to be expected in adapting it to commercial operations.

As early as 1947 Mr. Steffensen set up some small pelletizing equipment in his laboratory and a short time later built a larger experimental furnace. By separating the fine material from the coarse Cornwall concentrate, he could make a product fine enough for pelletizing and a coarser product that was more satisfactory for sintering. Although communication among the various firms was so poor at the time that we did not know it until 1951, Mr. Steffensen by 1950 was successfully operating a commercial furnace capable of producing 300 tons of pellets a day. His furnace shaft was rectangular, five feet wide and ten feet long. It was equipped with combustion chambers, chunk breakers, and good heat recuperators, and the fuel consumption was low — the equivalent of something less than fifty pounds of coal per ton of pellets produced.

Although he was not working with taconite, Mr. Steffensen and his engineers made a considerable contribution to the Minnesota taconite program. In 1950 both the Erie and Enterprise experimental pelletizing plants were having trouble conquering the problems of ball breakage and the discharge of clinkers from their furnaces. I saw Mr. Steffensen at a technical society meeting and inquired how he was progressing. "Everything's fine," he replied. "We're making good pellets in continuous operation." At his invitation, I went out to see his plant. There I found that he had solved the clinker problem by putting oscillating, toothed rollers at the base of his furnace to break up the chunks. In addition, he had solved the problem of ball breakage by adding a small amount of bentonite to the balling drum feed to make the balls tougher.[18]

Bentonite is mined in Wyoming and other western states as a soft rock or a hard clay. After being dried and crushed to a fine powder, it will absorb moisture and swell greatly. Just why a few pounds of it — 15 or 20 per ton of taconite concentrate — make the balls tougher and able to withstand rougher treatment without breaking is not completely understood. Bentonite also gives the balls another valuable characteristic. For some still unexplained reason, damp balls containing it can be heated to a high temperature much more rapidly without breaking or exploding. Both of the qualities that bentonite provides are unimportant in small-scale laboratory work, but they become very important where rapid firing and rough handling are necessary in commercial operations. Many other materials, such as lime, soda ash, and starch, have been investigated, but up to 1964 bentonite has proved to be the cheapest and most satisfactory additive.

Although Mr. Steffensen had found solutions for two of the big problems that were plaguing taconite research in 1950 and 1951, Bethlehem was not particularly generous about giving out information. I was elated to find that he had solved the problems that stalled the rest of us, but I did not feel free to reveal Mr. Steffensen's success. So I asked if he would be willing to address a forthcoming seminar on ore problems at Duluth. He could not go himself, but he agreed to send one of his men, Bernard Larpenteur, to comment on a paper which Mr. Wade and I presented. Mr. Larpenteur, a former Mines Experiment Station staff member, told the group about the first completely successful operation of a commercial pelletizing furnace.[19]

Much credit is due Mr. Steffenson and his technical staff — Mr. Larpenteur, A. F. Peterson, Jr., and T. R. Fredrickson — for their timely contributions to the development of commercial pelletizing. If they had not demonstrated that pelletizing could be carried out successfully with large-capacity equipment under commercial conditions, the whole taconite program might have been deferred or even abandoned.

As things stood in 1951, Bethlehem was well advanced in the development of commercial pelletizing, and three other companies were seriously investigating the concentration of Mesabi magnetic taconite — Erie, Oliver, and Reserve. We had come a long way in the ten years since the possibilities of taconite were first described to the people of the range communities and to the Minnesota legislature, but there was still a hard path to be traveled before the world was to have its first large commercial taconite processing and pelletizing plant.

7 The Crucial Years

THE CLOSING YEARS OF THE 1940S were crucial ones for Minnesota taconite. Would the firms showing an interest in this material move ahead? Could the remaining bugs in the process be ironed out quickly enough? From 1944 to 1951 the future of taconite hung in the balance. Those of us who believed that Minnesota's future importance as an iron ore producer was also being weighed held our breath awaiting the outcome.

While the struggle to perfect taconite processing was going on at the Mines Experiment Station and elsewhere, Reserve was taking its first steps toward commercialization. From 1944 through 1951 the firm moved forward, one tentative step at a time, toward the building of a large-scale taconite plant. In succeeding pages, we shall trace the Reserve Mining Company's program in greater detail than those of the other companies. It was with this firm that the author worked most closely and which was to build — following the Mines Experiment Station's recommendations — the world's first large taconite processing and pelletizing plant.

Since I had never before been associated with such a large venture, the process seemed surprisingly involved to me. Like most nonbusinessmen, I suppose I thought companies simply made up their minds to build plants and went out and built them. It was not that simple! Nor was it that irrevocable. At any point in the complicated process, it was possible for the company to change its mind. Seven years were to pass before Reserve committed itself to Minnesota taconite—and even then it could have backed out.

The company made its first definite move toward commercialization in 1944, when it employed I. H. Wynne, a skillful and imaginative industrial designer, and established an office in Milwaukee to formulate preliminary plans for a plant that could ultimately produce 10,000,000 tons of pellets per year. Reserve did not expect to build a plant of this size initially, but plans were made and space allocated for the various units that would be needed if and when the project was expanded. Under Wynne's direction, the long and frustrating task of transforming a laboratory process into a big commercial operation got seriously under way. In working out the plans, the designers were to follow exactly the improved flow sheet we had developed at the Mines Experiment Station. As soon as Wynne and his men understood the process, they began to encounter problems that had never crossed our minds even in our worst nightmares.

For one thing, when the designers realized what was involved in producing 64 per cent iron concentrate using the university's process, they concluded that there was not enough water available near the mine at Babbitt for so large a plant. Babbitt is exactly on the divide between the St. Lawrence and Hudson Bay watersheds, and the ridgepole of a roof is a poor place to collect water. The Mesabi Iron Company had faced this same problem some twenty-five years before, but its processing methods and the smaller scope of its project did not require nearly as much water as did the new flow sheet. The last washing step alone would take about 15 tons of clean water for each ton of concentrate produced. The other grinding and concentrating steps required about 30 tons, making a total of 45 tons of water for each ton of concentrate.

The 30 tons could be somewhat cloudy, recycled water—that is, water which had been used once, clarified for several weeks in a settling basin, and then returned to the plant to be used again. Such a settling basin would have to be very large and deep, so that waves would not stir up the silt after it had settled. There was no place available at Babbitt for such a big pond.

It quickly became apparent that water supply and tailings disposal were closely related and potentially serious problems in a project the size Reserve was considering. An installation to turn out 10,000,000 tons of taconite concentrate per year would also produce about 20,000,000 tons of tailings or sandy solids. These would be discharged from the plant with water, the amount of water being about twenty times the weight of the tailings. At 17 cubic feet per ton, the tailings would amount to over 340,000,000 cubic feet—enough sand to cover 40 acres nearly 200 feet deep each year. This meant that the tailings reservoir also would have to be very large just to hold the sandy solids discharged from a big taconite plant during its expected life of 25 to 50 years.

In 1945 and 1946 Oglebay Norton and Company sent Henry Martin and others to visit large copper producers in Arizona, Utah, and Nevada, where water was scarce, to study methods of tailings disposal. At the western plants a system is used of building dikes of the fast-settling coarser tailings around the edges of a pond and flowing the finer tailings and the water toward the center; the water is pumped back for re-use after the solids have settled. As the pond fills up with tailings, the dikes are raised and the pond is again filled. When it reaches its maximum safe level, another spot is selected. In the end a very large area may be deeply covered with fine sand. In the arid districts of the West, where there is little vegetation and a sparse population, this is satisfactory, but in a more populated area like Minnesota, dust storms from such a large sandy area could be very unpleasant. For this reason plans would have to be made to fertilize and irrigate, or perhaps cover the sand with topsoil to promote the growth of vegetation and prevent the sand from blowing.

In view of these problems, Wynne and his engineers conceived the idea that Babbitt taconite should be shipped to Lake Superior to be concentrated and pelletized. They reasoned that such a plan might solve not only the problems of water supply and tailings disposal but also reduce the cost of electric power, because a generating plant could be built at the lake and coal could be delivered to it in the ore boats that came to get the pellets.

While this idea seemingly solved three big problems, it created about that many more. With the mine located at Babbitt and the plant on the lake shore, a railroad some fifty miles long would have to be built to connect the two. Moreover, a spot would have to be found on the lake which was not only suitable for a harbor, but which also offered sufficient space for the plant and houses for its employees. Although this would mean a much larger investment for a power plant, a railroad, harbor facilities, an ore dock, and a town, the designers' figures indicated that such a plan would actually be the most economical in the end. They calculated that for each ton of pellets produced, the railroad would have to haul three tons of taconite from the mine to the plant. They pointed out, however, that a private interplant railroad could do this quite economically.

At about this point in the planning, the magnitude of the Reserve project began to register on me. The company would use more water and more electric power than the city of Duluth, and its railroad would transport to the lake more ore annually than any existing railroad carried to Duluth. Over 2,500 men would be employed at Reserve's mine and in its plants on a year-round basis. These men would be paid more than $1,500,000 a month, and towns to house them and their families would have to be built. Although the entire project would not spring into being overnight, it was hard to realize what all this could eventually mean to northern Minnesota.

But any plan to erect a plant on the shores of Lake Superior faced two hurdles: before it could be built permission must be secured from federal and state authorities to construct a harbor and to use water from Lake Superior. The lake does not wear its name idly. The largest and deepest of the Great Lakes, this inland sea is a magnificent body of cold, fresh water. Reserve did not wish to mar its beauty. If the firm erected a plant on its shores, it would have to obtain permission to take water from the lake and return it with sandy tailings from the taconite process. Before making such a request, the company wished to know what effect this would have on Lake Superior.

To find out, we made a series of tests at the Mines Experiment Station and at the St. Anthony Falls Hydraulic Laboratory of the University of Minnesota. To measure the settling rate of the tailings and to gauge the effect of adding them to the lake in various ways and under various conditions, a number of small tank tests were made at the station. They were repeated and expanded by Professor Straub using larger models in the hydraulic laboratory. The tests demonstrated that fine tailings entering a large deep body of water form what is called a "density current," which follows the bottom of the lake to the deepest point. The coarse tailings settle immediately and form a delta near the shore, but the fine tailings, as a dense current, flow to the bottom of the lake without coming to the surface. The deep water, which is not disturbed by wave action and surface currents, clarifies itself in a few weeks.[1]

As a result of these investigations, we at the Mines Experiment Station concluded that the best place for fine taconite tailings was in the deep valleys at the bottom of Lake Superior.[2] There they would be out of sight forever and posterity would not have to cope with them. We assured Reserve that the gray, sandy tailings of magnetic taconite would not in any way pollute the lake, interfere with any domestic water supply or with navigation, and would not adversely affect the fishing industry. It was our conclusion that the fine tailings from all the magnetic taconite on the Mesabi could be put into the deep water of Lake Superior and would have no harmful effect on its usefulness or beauty.

With this assurance Oglebay Norton, as Reserve's agent, began in 1945 to acquire land for a plant site a few miles east of Beaver Bay on the north shore of Lake Superior, where two small islands a short distance offshore could be used as anchor points for breakwaters to enclose a harbor. It was also decided to acquire additional land for a town to house employees and for a railroad right of way between the proposed plant location on the lake and Babbitt.

Lloyd K. Johnson of Duluth was employed by Oglebay Norton to buy the required land. The Northern Land Company and the Lake Superior Land Company were organized for the purpose of acquiring and holding the property until it would be needed by Reserve. Mr. Johnson was not permitted to tell the people of Beaver Bay and the surrounding area — if he knew — the purpose for which the land was being purchased. He was permitted, however, to pay prices well above the going rates, and this convinced the residents that some big, well-financed organization had important plans for the district.

Many were the stories that circulated through the old town of Beaver Bay, where seventy-five years before Peter Mitchell may have started his cross-country trek to discover the very taconite rock that Reserve was now planning to utilize. Moreover, Reserve's proposed railroad right of way pretty well followed the old trail that Mitchell and the Wielands had probably taken to Birch Lake. The 1945 land boom was the third that had occurred in the settlement. The first took place from 1855 to 1866 when the land was opened for homesteading and when copper, iron, gold, and other minerals were thought to have been discovered in the area. The second was in the 1880s when it was rumored that the Duluth and Iron Range Railroad would run from Vermilion Lake to Beaver Bay. Now in 1945 another boom was under way. Along Lake Superior's north shore and in the country back of it large blocks of land were being bought and paid for in cash.

Lorntson's store and lunchroom at Beaver Bay was the bus stop and the gathering place for people living in the area. As news of the sale of each new piece of land reached the store, more and more rumors and theories developed to explain the purchases. The local people reasoned that the pine and most of the pulpwood were gone, so no large development could be expected in the timber industry. They said that copper was not plentiful on the north shore and that the iron in the immediate area occurred only in small pockets and contained titanium, an element that made it undesirable for steel production. They reminded each other that the gold of the north country had turned out to be a myth, and so, they reasoned, the only remaining attraction was the tourist business.

Gradually the notion circulated that the land was being acquired for a large hotel and resort to be located just a few miles east of Beaver Bay near a bus stop called Silver Bay. Then it became known that Mr. Johnson had purchased the two rocky islands in Silver Bay and that sounding crews were measuring the depth of the water. The local residents believed this to mean that a harbor was planned for small boats bringing customers to the hotel. They assumed that the land being bought back from the shore was to be used for a golf course and a ski slide.

Up to 1946 only aerial surveys had been made by the Aero Service Corporation of Philadelphia, but in the summer of 1946, when ground crews appeared in the area, it became more and more difficult to keep the true facts from becoming known. It was also obvious that as soon as the real reason for the purchases was announced the price of land between Beaver Bay and Babbitt would increase enormously, and property along the north shore would take on new and greater speculative values. Before the ground crews appeared to make preliminary surveys for the railroad between Babbitt and the lake, there was not the slightest rumor, so far as I know, that Reserve was considering a large-scale project.

In October, 1946, at the earnest solicitation of some of us at the university, R. C. Allen (who was then president of Reserve) released the news that the company proposed to build a taconite plant near Beaver Bay and connect it by railroad with the mine at Babbitt. The issue of *Skillings Mining Review* for October 12 said: "The company has under consideration the use of these lands for a harbor and plant site in connection with its contemplated operations in taconite on the east Mesabi range. The Reserve Mining Co. has not made definite plans at this time just what portion of the plant, if any, shall be on the lake shore. . . . Should the mine rock be transported to the lake shore, the mining company will need a railroad, but if crushed rock or concentrates are transported to the lake shore other means may be adequate. The above is still under study by the Reserve Mining Co."

The announcement of the company's plans, of course, ended all rumors. At last after years of waiting and several disappointments Beaver Bay was to get a new lease on life. When I passed that way in the fall of 1946 on a hunting trip, Arthur Lorntson, the community leader who was later the village's first mayor, was cautiously jubilant. Although he felt that its simple quiet way of life would probably be disturbed, he predicted great prosperity for the little community.

Early in 1947 Oglebay Norton's attorneys began the long task of securing for Reserve the necessary permits from agencies of the federal and state governments to build and operate a taconite processing plant near Beaver Bay. In all, about a dozen permits were required for such items as constructing a harbor, erecting a power plant with a high smokestack, diverting creeks, altering slightly the location of Highway 61, crossing this highway with a railroad track, tunneling under it, locating other rail crossings on the line between the plant and the mine at Babbitt, drawing water from Lake Superior for use by the people of a new town and returning it after sewage treatment, withdrawing water from the lake to wash the taconite, and returning it to the lake with sandy tailings. With each application, detailed plans and drawings had to be submitted.

As it worked out most of the permits were granted without delay. Two of them, however, were in process for almost a full year. The Minnesota Department of Conservation would rule on taking water out of the lake, and the Minnesota Water Pollution Control Commission would decide whether the water and tailings could be put back into it. These agencies held a series of public hearings before ruling on Reserve's applications, which were filed on January 28, 1947. Chester S. Wilson was both commissioner of conservation and chairman of the water pollution control commission, and he presided at nine hearings held between June and December, 1947.[3]

At the first one, which was held in Two Harbors on June 5, 1947, the star witnesses were Harrie S. Taylor, general mines manager of Oglebay Norton, and Adolph G. Meyer, a Minneapolis consulting hydraulic engineer who had formerly been on the faculty of the University of Minnesota's College of Engineering. Taylor explained the whole Reserve plan—just about as it later developed—using charts and drawings and giving figures on cost, employment, and payrolls. Meyer and I discussed the technical problems of taconite processing and of withdrawing water from Lake Superior and returning it to the lake with tailings. We described the sandy tailings and explained the settling rates of the particles of various sizes. Speaking for the university, I predicted that the tailings would have no lastingly harmful effect on Lake Superior.

Meyer went into the results of the experiments conducted at the university, which showed that when taconite tailings flow into a large tank partially filled with water, the sandy portion of the tailings settles immediately to form a delta or beach near the point of discharge. The fine silt flows over this delta into the deeper water of the tank and then flows along the bottom to the lowest point, just as it would if the tank contained no water. In other words, the fine tailings in suspension act like a liquid that is heavier or denser than pure water and therefore flow immediately to the lowest point.

At the September 30 hearing, Dr. Straub, director of the university's hydraulic laboratory, explained that this phenomenon is frequently observed in the extensive reservoirs formed by large dams. Among the publications he cited was one illustrating the density current that forms where the muddy Colorado River enters Lake Mead behind Hoover Dam in Nevada.[4] The high-density muddy water of the Colorado, which annually carries much more silt than Reserve proposed to put into Lake Superior, goes immediately to the bottom of the reservoir, leaving the upper water clear and clean. Dr. Straub predicted that the same pattern would be followed if taconite tailings were introduced into Lake Superior.

As the hearings progressed a good deal of testimony was introduced favoring the granting of the permit to Reserve. Resolutions were presented on behalf of the city of Two Harbors, the Lake County Civic Association, the town board of Crystal Bay, and a group of resort owners on the north shore. Others who spoke were representatives of the Duluth Chamber of Commerce; C. A. Dahle, state senator from Duluth; George Laing, secretary of the Minnesota chapter of the Izaak Walton League; L. G. Rudstrom of Little Marais; Fred E. Wedell, Two Harbors city attorney; George Lyse of Crystal Bay; and John M. Jacobson of Beaver Bay. Wisconsin conservation officials who attended the hearings indicated they had no objection to the granting of the permits, and members of the United States Fish and Wildlife Service said the same.

The strongest and about the only opposition to the proposal came from representatives of the West Duluth chapter of the United Northern Sportsmen Club, Duluth locals of the Brotherhood of Railway Trainmen, the United Steelworkers Union (Congress of Industrial Organizations), and the Brotherhood of Railway Conductors. Several north shore fishermen also registered their objections, and state Senator Homer Carr of Proctor "proposed that the company dump the tailings on to the ground, with the state reimbursing it for the difference in cost from the water disposal system." [5] He offered to introduce such a bill in the next session of the Minnesota legislature. At the third hearing, Duluth public health officials asked whether the tailings would affect that city's water supply. But most of the opposition centered on the contention that the introduction of tailings into the lake would be detrimental to its fish life.

On July 22 Wilson granted opponents of the measure an extension until September 4 to prepare and present their cases. At the September session representatives of the United Sportsmen group and of the unions mentioned above introduced ten hours of testimony—some of which got rather outlandish. One of their witnesses maintained that the fine taconite tailings "would be carried out into the lake" where they would "stay in suspension," and the "farther down you go into water the slower those tailings would settle." "We know," he went on, "that water has a weight of 27 inches to a pound, and as you go down deeper it multiplies one pound every 27 inches." From this he reasoned that "an object will sink a certain distance in the water and then stay suspended." Any high school student could explain why this was not true. The answer given at the hearing was that if it were true, Lake Superior would be nothing but a big mudhole since existing streams had been pouring muddy water into the lake for thousands of years.[6]

Dr. John G. Moyle, aquatic biologist of the Minnesota Department of Conservation, brought out that tests "indicated the dumping of tailings into the lake would have no harmful effects upon its fish life." [7] He noted that fishing had not been good along the north shore for some years and pointed out that a rocky lake bottom does not furnish good spawning grounds for fish. It became obvious that the opposition was trying to drag out the hearings and delay the construction of a taconite plant on the lake shore. Why they were doing this was never clear to me.

At any rate, Chester Wilson and the authorities in St. Paul eventually terminated the hearings on November 4. Next day the *Duluth News-Tribune* said in summary: "Opponents of the plan . . . have maintained the tailings would tend to pollute the lake, with possible resultant effects to both fish life and humans. Experts, however, testified that the tailings were sand particles which would quickly sink to the bottom in the immediate area of the installation."

On December 18 and 22, 1947, the state authorities granted Reserve permits to take 130,000 gallons of water a minute from the lake and return it with the taconite tailings. The permits allowed the company a zone, extending three miles along the lake shore and three miles out into the lake, outside of which the discharge of tailings was not "to result in any material clouding or discoloration of the water at the surface," except when storm waves stirred up the mud in the shallow water, which is not unusual along the north shore.[8]

In Wilson's summary of the testimony presented at the hearings he stated: "The plans of the applicant for said project provide for the most practical and practicable use of the waters of the state, in pursuance of the policy of the state for the conservation of its water resources in the general public interest as declared by law, and will adequately protect public safety and promote the public welfare, and said project will be in furtherance of the best interests of the public and will not be materially detrimental to any public interests."

As a result of the hearings, Reserve's plans were pretty thoroughly aired. In some Duluth mining circles they were greeted with great skepticism by men who thought it highly unlikely that taconite could compete with the natural ores remaining in the state. They felt that "a bunch of outsiders" (Oglebay Norton had practically no operations in Minnesota) were moving into the iron country for a second time and trying to tell the experienced miners there what to do. They said that these newcomers did not even know as much about low-grade ores as Jackling, and his company went broke on taconite, that the theorists at the university had misled Oglebay Norton, and that poor old Crispin Oglebay had been taken in by them.

Not all the taconite skeptics lived in Minnesota. During this period when taconite's future hung in the balance, public interest was high. Newspaper and magazine articles, both technical and popular, were appearing, and the taconite project was discussed in Congress.[9] Marvin J. Barloon, professor of business and economics in Western Reserve University at Cleveland, published a widely read article in *Harper's Magazine* of August, 1947, in which he predicted dire results if the steel industry should put its reliance on taconite. Among other things, he said that the "added cost of ore made from taconite, by present processes, would amount to nearly five dollars a year for every American family." He did not explain how he arrived at the figure, but he concluded that foreign ore, not taconite, was the answer to the future supply of the nation and that the steel industry would move to the seacoasts where foreign ores were more readily available. The idea that steel might migrate because of a shortage of domestic ore was also used by proponents of the deep St. Lawrence Seaway to indicate that the waterway was necessary to preserve the industry in its present location by bringing foreign ore to the Great Lakes steel centers.

While these serious questions were being discussed, we at the Mines Experiment Station were not idle. There, as well as in the various test plants that had been erected, much attention was being given to pellet production. No pellets had ever been smelted, however, as any major proportion of a blast furnace burden. Many tests had been made on them for reducibility, hardness, strength, and resistance to abrasion, both hot and cold, under oxidizing and reducing conditions, but no actual smelting test had been made with pellets alone. The Mines Experiment Station had a small blast furnace and had accumulated some 120 tons of taconite pellets of indifferent quality during its experimental investigations. Logically a smelting test seemed to be the next step.

With the aid of Frank E. Vigor, vice-president in charge of operations for Armco Steel Corporation, at Middletown, Ohio, an invitation was conveyed to Armco to send some experienced blast furnace operators to the station to smelt pellets. Our furnace had a three-foot hearth, was about thirty-five feet high, and had a capacity when smelting standard ores of about five tons of pig iron per day. The university agreed to supply the ore, coke, and limestone required, maintain and operate the mechanical equipment, and furnish miscellaneous student help. Armco agreed to send experienced personnel to supervise, direct, and actually operate the furnace. It was expected that it would be "blown in" on standard ore and, when everything was operating well, pellets would be added in increasing amounts until the furnace was operating wholly on pellets. The test, which was to last about a month, would give the operators an opportunity to decide for themselves whether or not taconite pellets smelted as well as we at the Mines Experiment Station thought they should.[10]

The furnace was first ignited on February 23, 1948, by James L. Morrill, president of the university, and the author. After several false starts, it finally went into smooth operation late in April. It was blown in on standard ore of a type that the Armco men had used in their furnaces at Ashland, Kentucky. Operations continued for about a week on this burden until normal tonnages and analyses were established, and until the operators were satisfied that the furnace responded like the larger ones with which they were familiar. Then the ore was gradually shifted to pellets; ten days later the burden was 100 per cent pellets. Smelting continued for another five days until the supply of pellets was exhausted on May 20. The technical results of this test were turned over to Armco and were not made public. The most generally convincing result, however, was the sale of about 150 tons of pig iron it produced. Armco and Reserve gave this pig iron to the university, and we sold it on the then operating "black market" for about $57.00 per ton, which we turned over to the institution's scholarship fund.[11]

We also sent some of the pig iron to the university's foundry, where it was cast into replicas of the "Lynn Pot." This pot was the first iron casting produced in the United States; it was made at Lynn, Massachusetts, in 1648 by the Saugus Iron Works. We secured a photograph and drawing of the pot and cast about a hundred replicas which were presented in May, 1948, to workmen, mining company officials, legislators, and others who had made contributions to the advancement of the taconite program.[12]

By this time, Oglebay Norton had done a fine job of putting the whole Reserve taconite project together on paper with preliminary plans, town, railroad, and harbor designs, and construction and operating cost estimates. The firm had acquired for Reserve leases on large quantities of taconite, had secured land for plants, two towns, and a railroad, and had obtained the necessary permits from state and federal agencies to complete the project. The one thing missing was an assured market for the pellets. Jackling's Mesabi Iron Company had made the mistake of trying to sell its product on the open ore market at the Lake Erie price. To be assured of success, Reserve should be owned by steel companies who themselves could consume all the pellets produced. Of the various owners of Reserve at the time, only Armco and Wheeling were consumers of ore. Another large steel producer was needed in the organization, and Crispin Oglebay and his associate, Harrie Taylor, set out to interest such a firm.

Oglebay Norton had made frequent progress reports to Reserve, and some of this information was made available to officers of various steel companies who were known to be interested in increasing their ore reserves. A complete and impressive project could be described to these officials — a Mesabi Range property from which a half billion tons or more of high-grade concentrate could be made, agglomerated, and delivered to lower Great Lakes ports as a blast furnace ore seemingly superior to anything previously smelted, plus all the land and other requirements necessary for a large operation. Although the investment would be large, it was expected that a loan could be secured which could be retired over the long operating life of the property. If the pellets were as good as we thought they were, the whole proposition was an attractive one.

At about the time Oglebay and Taylor were trying to interest steel companies in Mesabi taconite, however, the large newly discovered ore deposits in Labrador were being offered to the very same firms, and the possibilities of a deep St. Lawrence Seaway to deliver these ores were being held up before their eyes. Wheeling and Cleveland-Cliffs, which had been participating in the Reserve venture, withdrew. Nevertheless, Oglebay and Taylor persisted. Crispin Oglebay died in 1949, but a year later his efforts paid off when Republic Steel Corporation joined the project. Republic acquired the right to purchase a half interest, and Armco bought the other half, so that Reserve was soon owned entirely by these two ore-consuming firms.[13]

This did not necessarily mean that Armco and Republic would definitely go ahead with Mesabi taconite. Their officers were interested in the potential value of the pellets as an economical smelting product, although this had not as yet been conclusively proved. Stepping out with a large investment in a new and untried process and with a product that had not been completely tested commercially was debatable, to say the least. Both firms were principally concerned about assuring a future ore supply for their big steel plants. Some of their officers regarded a domestic ore reserve as preferable to a foreign one, but both companies also acquired options in the Labrador ore fields. Whether they finally obtained their ore from Minnesota taconite or from Labrador remained to be seen.

We at the Mines Experiment Station did what we could to encourage a vote for the pellets. To acquaint the officials of Republic with the simplicity of taconite processing, a complete demonstration from the crude rock to the finished pellets was arranged for November 16, 1950, at the Mines Experiment Station. Planes brought the men to Minneapolis from Cleveland and Middletown early enough in the morning so that the two groups reached the university shortly after 10:00 A.M. Our equipment had been assembled in the laboratory so that taconite rock from Babbitt could be fed in at one end, and the final concentrate would be discharged at the other. The high-grade concentrate was then taken to the pelletizing furnace, and pellets were produced at the rate of one ton per hour. For once everything worked smoothly.

The men watched the demonstration until noon, and then we all went to the Campus Club in the Coffman Memorial Union for lunch. President Morrill, who had been at Ohio State University before going to Minnesota and had known some of the Armco people, came in briefly to greet the guests. At the table were: Charles M. White, William M. Kelley, Robert J. Linney, Edward B. Winning, Oscar Lee, and Fred M. Darner of Republic; Henry Martin of Oglebay Norton; Kenneth C. McCutcheon and W. Edward Marshall of Armco; Vice-president William T. Middlebrook, Professor Robert T. Jones, Henry Wade, the author, and a few others from the university.

Dale D. Shephard, manager of the Campus Club, put on a fine luncheon for us. He had sent some of his boys out to shoot enough pheasants for the group. Afterward Mr. Kelley liked to tell how the university bribed the visitors with pheasants to get them into the taconite business. Each time he told the story the pheasants got larger until they were as big as turkeys.

In the afternoon we went to the university's department of architecture where Professor Jones and his graduate students had constructed three complete models of the proposed town which eventually became Silver Bay. Then we returned to the Mines Experiment Station so the visitors could watch more of the pelletizing operation. Although the equipment these men saw in our laboratory was small and some of the operations were awkward, they could visualize the process improved and enlarged. In general, they felt that the various steps would not be difficult to commercialize.

Charles White, who was president of both Reserve and Republic at the time, had once been a blast furnace operator. It was he who a few years earlier had told me that he wanted no part of tax-hungry Minnesota and its taconite. At the station he saw the pellets in a pile for the first time. He picked up a handful and said to me, "I'm afraid these damned pellets of yours are going to get me into the Minnesota taconite business." Seeing the round pellets in a large pile, Mr. White realized that, because of their shape and strength, they would not pack down in a blast furnace to form a tight, impervious mass. No matter how they might be charged, or how they might arrange themselves, the furnace gases would always have free access to every pellet. Mr. White's recognition of the superior physical quality of the pellets was of great importance, since it would probably be he, more than any other single man, who would influence the final policy decision on whether Reserve would begin large-scale commercial taconite processing in Minnesota.

Later other visits were made to the station by Republic and Armco designers and experts. At last, in September, 1951, Reserve announced that it would go into the taconite business in a big way. Its program, following the plan proposed by Oglebay Norton, was to open a mine, build a crushing plant, and establish a new town at Babbitt. A railroad was to be constructed from Babbitt to Silver Bay on Lake Superior. There the initial concentrating and pelletizing plants would be built—with a capacity of 3,750,000 tons—as well as a new harbor and loading dock, and another town. But the first step would be the rehabilitation of the old Mesabi Iron Company's Babbitt plant to test the new flow sheet and pelletizing process developed by the Mines Experiment Station.[14]

The decision of Republic and Armco to go ahead with Minnesota taconite was the break-through we had been working for so long. Serious discussions, led by Presidents White and W. W. Sebald, preceded the decision of the boards of directors of Republic and Armco. The superior smelting qualities that had been designed into the pellets and the willingness of these men to gamble on a possibly vast but as yet unproven domestic ore supply apparently won the day. But it was a close and difficult choice. In the face of cheap foreign ore and a new St. Lawrence Seaway, it took courage to decide in favor of the hard, tough, low-grade taconite on the eastern Mesabi. The men who voted their companies into this untried and expensive process, fresh from a university laboratory, were laying their careers, as well as the futures of their firms, on the line.

Mr. White once introduced me to his board of directors by saying, "Gentlemen, I want you to meet the man who sucked us into this taconite thing." I am convinced, however, that it was Mr. White's look at that pile of pellets on the floor of the Mines Experiment Station's laboratory in 1950 that really got him into the Minnesota taconite business. Whatever the cause, Mr. White's enthusiasm for the pellets, his dynamic personality, and his record of successful ventures in the steel industry was to be largely responsible for the building of the world's first big commercial taconite plant. Doubtless many factors contributed to the decision of Republic and Armco — Mr. White's attitude and the Minnesota taconite law being by no means the least — although Mr. Kelley always maintained that it was those big Minnesota pheasants which tipped the scales in favor of taconite. We never told him the pheasants came from South Dakota.

8 The Next Step

By the early 1950s many of the questions about taconite had been answered affirmatively, and the program was definitely moving forward. Not only was Reserve taking the next step toward large-scale taconite production but Erie and Oliver were also active.

After opening its laboratory in West Duluth in 1944, Oliver—like Erie—at first went its own secretive way. Walter L. Maxson, a graduate of Cornell University and the Colorado School of Mines, was in charge of the Duluth research staff. He had previously been employed by the Allis-Chalmers company as an expert on the design and operation of rod and ball mills and other mineral-processing equipment manufactured by that firm. In 1949 Oliver field crews began extensive drilling of the company's taconite properties and the Duluth laboratory tested the many samples submitted. The company reported that "five miles of test holes were drilled" before a mine site was chosen near Mountain Iron—the town founded by the Merritt brothers when they opened the Mesabi Range in the early 1890s.[1]

In 1950 ground was broken near Virginia for an experimental agglomeration plant, and in May, 1951, the firm began the development of a taconite mine and the construction of a pilot concentration plant near Mountain Iron. The concentrating plant, which the company named "Pilotac," was to have a capacity of 500,000 tons per year. The concentrate was to be shipped some

CRISPIN OGLEBAY (ABOVE) AND CHARLES R. HOOK (BELOW) *were the chief executives of Oglebay Norton and Company and American Rolling Mill Company (now Armco) when Reserve was organized in 1939. The steel men called the new firm "Reserve" because they did not expect to have any immediate need for taconite. Courtesy Oglebay Norton and Armco.*

ROLLAND C. ALLEN *(left), Henry K. Martin, and Harrie S. Taylor of Oglebay Norton played important roles during the formative years of the Reserve Mining Company in the 1940s and early 1950s. Courtesy Oglebay Norton.*

WALTER L. MAXSON *was director of research in Oliver's Duluth laboratory in 1944. Courtesy United States Steel Corporation.*

PERCY L. STEFFENSEN *of Bethlehem Steel solved two big problems that were holding up taconite research in the late 1940s. Courtesy Bethlehem Steel Corporation.*

THE EXPERIMENTAL PLANT *of Iron Ore Pelletizing Enterprise at Ashland, Kentucky, was set up to develop commercial pelletizing equipment for Reserve. It operated from 1949 to 1952 and provided information which was later useful when Reserve's big plant was built at Silver Bay on Lake Superior.*

KENNETH M. HALEY, *who was in charge of the pelletizing plant at Ashland, later joined Reserve's staff at Babbitt and Silver Bay.*

CHARLES M. WHITE (ABOVE) AND W. W. SEBALD (BELOW) *were in 1950 the presidents of Republic and Armco, the steel companies which then owned Reserve. Their enthusiasm for taconite pellets influenced the decision authorizing Reserve to build the world's first large taconite concentrating and pelletizing plant at Silver Bay. Courtesy Republic and Armco Steel Corporations.*

AUTHOR E. W. DAVIS *(left), Minnesota's "Mr. Taconite," and William M. (Doc) Kelley, who was president of Reserve from 1954 to 1958.*

C. L. KINGSBURY (LEFT) AND OSCAR LEE (RIGHT) *were active members of the team that built Reserve's plants and two new Minnesota towns in the 1950s.*

FRED M. DARNER *designed the large taconite plant which was built by Reserve at Silver Bay on Lake Superior between 1951 and 1955.*

J. WILLIAM BRYANT, *who was Reserve's comptroller during the crucial building period, later succeeded Mr. Linney as the company's president.*

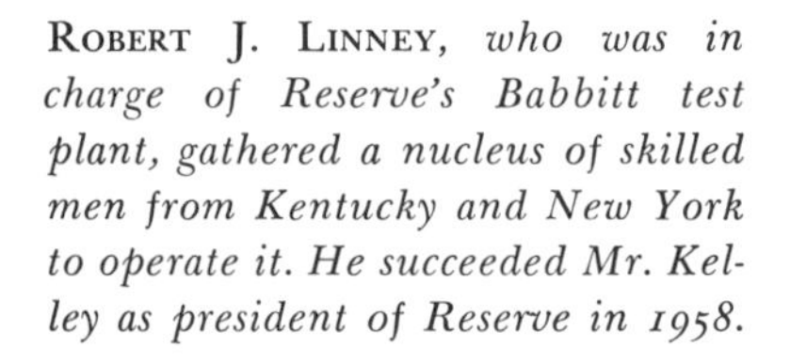

ROBERT J. LINNEY, *who was in charge of Reserve's Babbitt test plant, gathered a nucleus of skilled men from Kentucky and New York to operate it. He succeeded Mr. Kelley as president of Reserve in 1958.*

ALEX D. CHISHOLM *(left) and John C. Metcalf (right) of Pickands Mather and Company were early taconite enthusiasts. In 1940 Pickands Mather, acting for four steel firms, organized Erie Mining Company. Fred D. DeVaney (center) headed Erie's taconite research program. Courtesy Pickands Mather.*

ELTON HOYT II, *president of Pickands Mather, was the moving force in the development of Erie's program for making commercial use of taconite. Courtesy Pickands Mather.*

THE AURORA PILOT PLANT *of the Erie Mining Company, built in 1948, was until 1952 the only Minnesota plant equipped with commercial-sized machines for processing taconite. It closed in 1954. Courtesy Erie Mining Company.*

ERIE'S BIG HOYT LAKES PLANT, *built under the supervision of Pickands Mather, began to ship pellets in 1957. It had a capacity of 7,500,000 tons of pellets per year and was the largest taconite plant yet erected. The pellets were used by the steel firms which own Erie. Courtesy Erie Mining Company.*

A NEW TOWN NAMED HOYT LAKES *(for Elton Hoyt II) was built by Erie in 1954–55 on Minnesota's eastern Mesabi Range to house employees of the firm's big taconite processing plant. Courtesy Erie Mining Company.*

PILOTAC, *the experimental taconite concentrating plant built by the Oliver Iron Mining Division of United States Steel, went into operation at Mountain Iron, Minnesota, in 1953. Courtesy United States Steel Corporation.*

THE EXTACA PLANT *of United States Steel, located at Virginia, Minnesota, was opened in 1954 to test various methods of agglomerating the concentrate produced at Pilotac. Courtesy United States Steel Corporation.*

four miles to the agglomerating plant, which the firm called "Extaca." Pilotac was equipped with crushers, rod and ball mills, magnetic separators, and dewatering equipment arranged for flexible operation. At Extaca large units were installed to test sintering and nodulizing methods of agglomeration, but no pelletizing equipment was provided.[2]

Meanwhile Erie was operating its pilot plant near Aurora, the only one in the state from 1948 until 1952 with commercial-size equipment. In 1949 this firm applied to the necessary state agencies for permits to use water from the Partridge River drainage area on the Mesabi Range in a commercial plant with a capacity of 7,500,000 tons—twice the size of Reserve's initial unit.[3] In 1952 Erie announced that it would proceed with the erection of a big plant near Aurora in a new town which was later named Hoyt Lakes for Elton Hoyt II, head of Pickands Mather and Company, Erie's agent.[4] The entire project was to cost over $300,000,000 supplied by the four iron and steel companies which own Erie and by the sale of first mortgage bonds to a group of insurance companies. Plans for the plant were to be prepared by the engineering staff of the Anaconda Copper Company. And so, after the passage of over thirty years, copper men once more returned to the Mesabi Range to tackle Minnesota taconite.[5]

Erie's program called for the construction, beginning in 1954, of a concentrating and pelletizing plant near its mine at the old town of Mesaba. The firm later decided to build a railroad 73 miles long from the plant to a new harbor and loading dock it would construct on Lake Superior near the town of Schroeder. (The ore-loading facilities there were named Taconite Harbor in 1955.) Unlike Reserve, however, Erie would ship over its tracks finished pellets rather than raw taconite rock. A power plant was to be built at the harbor to supply the mine area over a new transmission line.

The plant in the new town near Aurora would be the largest concentrator thus far built in a single initial unit. It was to be so constructed that it could later be expanded to produce 10,500,000 tons of pellets annually. Pickands Mather was to supervise and manage the big design and construction job. The general features of the Erie project were much the same as those of the previously described Reserve project. Both were designed to produce pellets assaying between 64 and 65 per cent iron, and about 8 per cent silica. But there were four principal points of difference.

1. The Erie plant was to be built inland. Tailings would be partially dewatered and then pumped to a large disposal area just over the hill north of the plant. The company conducted experiments with the University of Minnesota department of agriculture to develop various grasses, shrubs, and trees to revegetate the area after it had been covered with tailings. Water for the plant would be recovered from the tailings supplemented by 12,000 gallons a minute from the Partridge River drainage area.

2. The grinding and concentration flow sheet would consist of one ball mill for each rod mill, 54 in all, with cobbers between. The rod mill would be fed at a lower rate than at the Reserve plant, thus making finer cobber tailings which could be dewatered and pumped more easily.

3. Pelletizing was to be done in 24 shaft-type furnaces similar to those developed at Lebanon by Bethlehem Steel Corporation but larger, 14 feet long, 6 feet wide, and having a cross section of 84 square feet.

4. The pellets would be shipped to Taconite Harbor in the summer only, and would be stock-piled in winter near the plant.[6]

All these announcements coming in rapid succession stimulated high hopes and visions of full employment in northern Minnesota, and a great upsurge of optimism developed in the early 1950s. Articles appeared in technical and popular publications under such titles as "Taconite Brings New Life to Minnesota's North Shore," "Taconites Beyond Taconites," "Where 1972's Iron Ore Is Coming From," "Taconite: Minnesota's Iron Storehouse for the Future," "Minnesota Taconite: Nature's Cinderella," and "The Professor's Crazy Billion-Dollar Dream."[7] Much remained to be done, however, before these great expectations could materialize. Plans can be announced, land acquired, permits secured, and preliminary work completed, but until adequate financing is arranged, there is no assurance that a taconite project will actually become a reality.

Reserve moved along more rapidly in building a large plant than did Erie. The company's staff was for the most part assembled by Presidents White of Republic and Sebald of Armco from their own organizations. Republic men were chiefly in charge of operations and Armco men were principally responsible for finances. Oglebay Norton continued to act as a consultant, and some of its men were transferred to Reserve, but management responsibilities were gradually shifted to the expanding Reserve organization. Mr. White was president; Mr. Kelley became vice-president of operations; C. L. Kingsbury, financial vice-president and treasurer; J. William Bryant, comptroller; Robert Linney, manager of operations; Fred Darner, executive chief engineer; and Oscar Lee, metallurgist. In 1951 the author took a leave of absence from the University of Minnesota to join Mr. Kelley's staff.[8]

William Myron (Doc) Kelley, who was to be the guiding spirit of the new venture, was a native of Pennsylvania whose formal education was interrupted at an early age because he had to help support his family. He went to work in a steel plant as an apprentice machinist and advanced rapidly to positions of higher responsibility, usually connected with the building, installing, and maintaining of steel plants. It was he who installed the equipment in the steel plant at Duluth and at Indiana Harbor on Lake Michigan. By 1944 he was assistant vice-president in charge of Republic's manufacturing division, and in 1949 he became vice-president in charge of all Republic's operations. A year later he was made vice-president of Reserve, and in 1954 he was to become that company's president at the age of sixty-five. The success of the whole taconite movement owes much to "Doc," as he was affectionately called.

I once said to Mr. White, "It's certainly fortunate that you have a man like Mr. Kelley to put in charge of this project."

Mr. White replied, "You have that just backward, Ed. If we didn't have a man like Doc to put in charge, we never would have gone into taconite."

Reserve decided to build a 3,750,000-ton plant on the shores of Lake Superior that could later be expanded to the 10,000,000 tons which had been discussed in the planning stages. Even so, arranging to finance such an undertaking was a big job. A total of $185,000,000 would be needed for the smaller unit. Republic and Armco, as the owners of Reserve, agreed to provide $37,000,000, or 20 per cent of the amount. An additional $148,000,000 was to be borrowed by issuing first mortgage bonds.

One does not simply run down to his neighborhood bank and get such a sum of money; in fact, one cannot expect to obtain it from any single source. As a first step, Reserve in 1952 asked me to prepare a report explaining the project in general terms. Such a report was necessary because taconite processing was new and commercially untried, and all the information potential lenders would have about it before they passed on this enormous loan would have to be provided. We wrote it assuming that the investment officers to whom it was to be delivered would know nothing about taconite or pellets and little about northern Minnesota and the Mesabi Range. We made it quite descriptive, covering the iron ore situation in general and Minnesota in particular, with a full discussion of the nature and occurrence of magnetic taconite, the earlier attempt that had been made to utilize it, the new Mines Experiment Station flow sheet, the nature of the plants, harbor, and towns to be constructed, and the quality of the pellets. The idea was to impress potential lenders with the fact that Reserve had the necessary raw material, the talent, and the ability to make a competitive product for which there was an assured market.[9]

As I remember the finished report, it was so voluminous that it cost $39.00 to send it via air express from Minneapolis to Cleveland. Some fifteen or twenty exhibits, maps, and photographs accompanied the text, and many of their original authors would have been surprised had they known to what use their statements were put. For example, a published address about taconite given before the Newcomen Society by President Morrill was included, as were many technical reports by experts in geology, metallurgy, hydraulics, water pollution, and taxation.[10]

Although the report offered a complete description of the whole project, I was told that it need not contain information on the cost of production or the dollar value of the pellets. This surprised me until I realized that Republic and Armco expected to take all the pellets Reserve produced at a price which would cover the total cost of production, the interest on the loan, and its amortization. Actually it was Republic and Armco, not Reserve, who were guaranteeing the mortgage and taking the risk. Any profit from the project would be made by Republic and Armco stockholders on the steel produced from the pellets. Reserve's job was to provide good pellets as cheaply as possible.

All this information was delivered to the investment firms of Glore, Forgan and Company, and Smith, Barney and Company, of New York City, and was presented by them to the loan departments of large banks and insurance companies known to have sufficient funds to participate in a $148,000,000 bond sale. After study and investigation, the institutions interested in participating indicated the portion of the bonds they were willing to buy. As it turned out, all the subscribers were insurance companies.[11]

But these companies did not turn over this money to Reserve all at once. From time to time each of them deposited a percentage of their subscription in the Chemical Bank and Trust Company of New York, and Reserve then drew on these funds as they were required. Reserve did not want to borrow the money and begin paying interest on it until it was actually needed. A large part of Bill Bryant's job as comptroller was to have the money available as Reserve's bills came due but not before. The interest on $185,000,000 at 4 per cent, for example, is over $20,000 a day, and until Reserve got its plant built and in operation it would have no income.

If the total amount of the principal were to be paid back in twenty-five years at 4 per cent, Reserve would have to pay nearly $7,500,000 a year plus interest, which would average about $3,750,000 per year over the twenty-five-year life of the loan. At an interest rate of 4 per cent, merely satisfying its financial obligations would cost Reserve over $11,000,000 a year. At the proposed production rate of 3,750,000 tons of pellets annually, almost $3.00 per ton of pellets would have to be added to the direct operating costs of labor, supplies, and so forth, in order to arrive at the total cost of the pellets.

If Reserve had not been able to negotiate a long-term loan, its annual commitments for principal and interest would probably have been impossibly high. For example, if the lenders had said they wanted their money back in ten years rather than twenty-five, the payments would have been $18,500,000 per year plus interest, adding an average of over $6.00 per ton to the direct operating cost of pellet production. This would probably have raised the price above that for which they could be sold. Thus the importance of a long-term loan becomes apparent, as does a long-time supply of taconite, some of which would not be mined and made into pellets for twenty years or more. These calculations illustrate the importance of the taconite tax law, under which a long-time supply of low-grade raw material can be held, and the taxes are collected, for the most part, not on the rock in the mine but on the high-grade pellets after they are produced and shipped.

Taconite producers have the opportunity of financing themselves in either of two ways. They can borrow the money or they can sell stock to the public. Up to the present time (1964), all taconite producers (except the early Mesabi Iron Company) have chosen to follow the traditional procedure in the development of iron ore properties in the Lake Superior district, and have borrowed the money to finance their plants. In that way the steel companies backing them take all the risks and retain complete control over the source of their raw material.

But to finance Reserve's initial venture, the possibility of forming a stock company must have been tempting. (Jackling had financed the Mesabi Iron Company that way.) For one thing the stockholders would then have assumed all the risk that pellets could be delivered to the steel companies at a price competitive with that of ore from other sources. If taconite processing had turned out to be uneconomical (and the only previous attempt had been a financial failure), the stockholders would have lost their money, and the steel companies would have been free to get their ore from other sources. Only if the project succeeded would the stockholders realize a return on their investment in the form of dividends.

Just why the steel firms backing the Reserve and Erie projects did not sell stock has not been explained. Perhaps they were advised that the public would not care to participate in such a new and untried venture. Many people now wonder why they were not given an opportunity to participate directly in this industry. If they had been asked in 1952—before any taconite pellets had been made or smelted commercially—would they have been interested in risking their money? Although it now seems to Minnesotans that Reserve and Erie are out-of-state corporations, in reality a considerable number of Minnesota residents have an indirect interest in these companies through their insurance holdings.

As soon as Reserve's financial program became known, amateur estimators in Duluth and elsewhere got out their pencils and began figuring. They had a few published figures to work with. The *Minneapolis Star* of December 26, 1952, announced that Reserve "today filed a mortgage covering its taconite properties" in St. Louis and Lake counties in the amount of $148,000,000. The amateur statisticians expected that some 2,000 men would be employed to produce 3,750,000 tons of pellets annually. Figuring an average of $3.00 per hour and a forty-hour week, they came up with an annual payroll of about $12,000,000, which would be about $3.20 per ton of pellets for labor.

Information released earlier by the Mines Experiment Station indicated that thirty pounds of coal would be used for each ton of pellets. At a delivered price of $10.00 a ton, the coal would cost about 15 cents for each ton of pellets produced. The Mines Experiment Station had also predicted that taconite processing would require a power consumption of about 85 kilowatt hours per ton of pellets. Using the Minnesota Power and Light Company taconite rate, the amateurs estimated that power should be available for about one cent per kilowatt hour, making a cost for power of about 85 cents per ton of pellets.

Some figures compiled in 1945 had indicated that the cost

of mining taconite would be about 30 cents per ton of rock or 90 cents per ton of pellets.[12] Coarse crushing, they figured, would add about 15 cents per ton of pellets for a total cost of $1.05. Since it would be necessary for Reserve to transport the three tons of crushed rock to Silver Bay, the amateurs estimated that this could be done at an approximate cost of perhaps $1.50 per ton of pellets. Adding these costs together, the amateur estimators got the following results per ton of pellets:

Labor	$3.20
Pelletizing Fuel	.15
Electric Power	.85
Mining and Crushing at Babbitt	1.05
Transportation to Silver Bay	1.50
Total	$6.75

To this it would be necessary to add the cost of taxes, royalties, insurance, social security, and so forth—which the amateurs estimated to be about $1.00 per ton—bringing the total direct operating expense to $7.75 for each ton of pellets. To this they added interest and amortization of about $3.00 per ton of pellets on a twenty-five-year loan, thus making their estimated total cost of pellets at the plant $10.75 per ton. Transportation to lower Great Lakes ports would add another $2.05, giving a total estimated cost of $12.80 per ton delivered. If the pellets assayed 63 per cent iron (natural) and could be sold as Mesabi Bessemer quality, the computed 1952 Lake Erie Base Price would be about $11.50 per ton delivered, or a loss of $1.30 a ton.[13]

Many Duluth mining men were openly skeptical of such a large and risky investment. As more and more information was released about the project, Reserve men received cooler and cooler receptions as they passed through Duluth. The ribbing at the Kitchi Gammi Club, a meeting place for the business and professional men of the area, became sharper and at times almost insulting. Oscar Lee, Reserve's metallurgist, often stopped overnight at the club on his trips from Cleveland to Babbitt. Late one evening he was invited to a farewell party there for one of The M. A. Hanna Company men, who was moving to the Cleveland office. The author was an interested spectator at this gathering of a dozen or so local mine operators. Some of them lost their inhibitions and said plainly that they thought Reserve was crazy, that without a single man in its organization familiar with Lake Superior mining, the firm was rushing into a new area with a completely untried process and spending money like a drunken sailor. Moreover, the Duluth or Two Harbors loading facilities were not good enough for Reserve which had to have a harbor of its own, and of all the just plain foolish things, the company was going to build its own railroad and haul three tons of taconite to the lake when it might just as well erect its plant at Babbitt and throw away two tons of tailings there before shipment. Then it would have only one ton of concentrate to haul, and that could easily be transported over a railroad which was already in existence.

Oscar had been around a bit and could take care of himself pretty well. He did not mince words. He told the Duluthians they had been living up there in the woods so long digging soft, high-grade ore and raking in the profits that they thought this was all there was to mining. They were, he said, resentful of anyone coming into their private mining area with anything new because it might interfere with their cocktail hours. He said it was about time they shook off their complacency and brought their thinking up to date. He told them that their railroads and ore-handling facilities were outdated and inefficient. He told them that Reserve's new harbor would be only a fraction of the size of the Duluth-Superior facilities, but in the end would ship more pellets than all the ore that had been shipped from Duluth and Superior put together since the discovery of the Mesabi. He told them that Reserve could not afford to use the old, obsolete, steam locomotives and small bottom-dump railroad cars they were using. Reserve's railroad would handle more ton miles of freight per year than any railroad in Minnesota. He told them that the days of putting raw ore, just as it was dug from the ground, into a blast furnace were about over, and that they had better start modernizing their thinking or they would be left behind.

When this encounter was reported to Doc Kelley, he brushed it aside. "Pay no attention to what they say," he counseled. "Our job is to make this plant operate at top efficiency. The pellets will speak for themselves." In this atmosphere, Reserve's men moved into the Babbitt area and began the work of setting up a test plant on the old Mesabi Iron Company site.

9 Babbitt Revived

WORK GOT UNDERWAY immediately on the mine, pilot plant, and new town at Babbitt in 1951.[1] Some of the equipment installed by the Mesabi Iron Company in the 1920s had been removed and sold, but much still remained. As a first step in converting the old buildings into a new test plant, the machines from the earlier venture, as well as most of the floors, supports, conveyer galleries, and bins had to be removed. Only the shells of the structures were retained. The new Mines Experiment Station flow sheet for concentrating and pelletizing at a capacity of about 300,000 tons of pellets a year was designed to fit into the existing buildings. This was really more difficult, although probably less expensive, than constructing a new test plant, principally because little new excavation and foundation work was required. For example, the old Mesabi crusher pit was eighty feet deep in solid rock, and it was utilized for the new crusher installation.

During the actual rebuilding of the test plant, Mr. Linney, assisted by Floyd W. Erickson, was in charge at Babbitt for Reserve. The designing and engineering were carried out by Gus Wynne, and the purchasing and accounting were done by Oglebay Norton at Duluth under the direction of Frank J. Smith.

The layout of the Babbitt mine, which was later appropriately and officially named for Peter Mitchell, received a great deal of attention.[2] Although initially it would be a rather small affair, there was nothing haphazard about its development. As it was planned, the mine area—which would furnish the 3,000 tons of rock needed per day by the Babbitt test plant—was eventually to be little more than an approach giving access to the big mine that would be needed after the large lakeside plant went into production.

The 3,750,000-ton unit at the lake would require over 30,000 tons of taconite a day, and it was anticipated that, if all went well, 80,000 to 100,000 tons a day would be needed by 1970. It was thought that the pit would eventually be enlarged until it was ten or more miles long, a mile or so wide, and several hundred feet deep. Thus the mine might well become one of the largest in the world within a few years, and its operation was carefully programed. Large models were prepared showing how the pit was to be opened, where each year's supply of taconite would be taken out, and where the access roads and power and water lines were to be located.

For the start of the mine, however, a location was selected just east of the old Babbitt pit. The taconite there was known to have good concentrating characteristics, and to be only thinly covered with overburden which would have to be removed to get at the rock. At first the blastholes would be shallow, as an inclined approach to the pit was formed. Later, working faces twenty-five to thirty-five feet high would be developed. A row of holes would be drilled about thirty-five feet deep in the solid taconite behind these faces, and the rock would be blasted down into the pit as it had been in the Mesabi Iron Company's operation.[3]

In contrast to the laborious churn drilling on that project in the 1920s, by the time we returned to Babbitt in 1951 a new and much faster drilling method had been developed by the Linde Air Products Company, assisted by Oglebay Norton. Both Erie and Reserve were experimenting with this new process, which was called jet piercing.[4] It literally burned holes into the rock with a machine using liquid oxygen and fuel oil to produce a temperature of about 4300 degrees Fahrenheit. It could drill a hole eight inches in diameter in the hard taconite at the rate of about thirty-five feet an hour. Since the blasting of each hole would produce about 2,000 tons of broken rock and the mining operation was being planned for an eventual production of more than 80,000 tons per day, it was evident that the drilling and loading of blastholes would be a big and continuous job, requiring many drill rigs, a large supply of liquid oxygen and fuel oil, and carloads of explosives—all of which must be delivered to the blasting area exactly as needed. Obviously this would take careful, detailed planning and scheduling.

Fortunately the mine could be developed slowly and its operation could be adjusted as experience was gained. For the first period—to supply the Babbitt plant with 3,000 tons per day—the blasted and broken taconite was to be picked up by electric shovels with five-cubic-yard dippers and loaded into twenty-two-ton trucks to be hauled about two miles over a new and carefully built road to the plant.

While the mine was being laid out and the Babbitt plant was being rebuilt, Bob Linney began to gather operating crews. Taconite processing was, of course, new to everyone, but various steps were in use at other plants. In his attempt to hire men who were familiar with some of these steps, Bob went first to Republic's magnetic properties in New York, where he had grown up and where he knew the capabilities of many of the men. From there he secured Edward M. Furness, as mill superintendent, Henry Genier to supervise magnetic separation, William Kowalowski to take charge of the fine crushers, and Bernard LaVigne, an expert on the maintenance of heavy equipment. These experienced workmen were the nucleus around which the locally hired operating crews in the concentrator were organized.

Bob then went to Ashland, Kentucky, to see whether he could persuade any of the key men in the Enterprise pelletizing plant to move to Babbitt. He expected trouble, because the company grapevine reported that these Southern workers were not interested in moving to cold, northern Minnesota. Bob not only had to persuade the men to take the jobs, but he also had to sell their wives on the idea of moving to Minnesota's frigid wilds. He took the entire group to dinner and told them stories about the invigorating climate, the beautiful white snow, and the comfortable, heated houses.

A couple of the Southern ladies wanted to know what the people did for entertainment in winter. Bob told them about skiing, skating, and ice fishing, but he did not seem to be making much impression until he began to describe the game of curling. This interested them. Later one of the ladies was overheard explaining that in Minnesota men and women played a game called "curling up" in a place especially built for it, and you could go and watch if you didn't want to play. She said it was sort of like an old New England bundling party. This was more like it, and Bob returned with four key men signed up: Ken Haley as pelletizing plant superintendent, Don Cooksey as his assistant, James Reynolds as chief operator, and Harold Trask as test engineer. Although Bob denies it, it must have been his description of "curling up" that took the sting out of the cold Minnesota weather. Eventually several more men were obtained from Ashland and from the Adirondacks, and it was the know-how of these experienced workers that made the Babbitt plant operate so smoothly.

To the families who moved to Babbitt from New York and Kentucky, northern Minnesota seemed as remote as the moon. They left their friends and relatives behind, and after the excitement of moving wore off, one or two of them—or their wives—couldn't take the remoteness and the frontier atmosphere and returned to their former homes. But most of them stayed and quickly adjusted to their new surroundings.

Babbitt was more isolated than Erie's plant near Aurora or Oliver's plants near Virginia. Since the closest town was Ely, some twenty miles to the northeast, it was necessary for Reserve to make immediate provision for housing its employees at Babbitt. The old town had been located on the bare rock of the Mesabi ridge, facing south toward Argo Lake. Within walking distance just over the ridge to the west were the mill buildings of the Mesabi Iron Company. The old town was not well situated by modern standards, and the available area was too small for Reserve's future expansion plans. Moreover, most of the existing structures were not adequate for permanent housing.

Reserve owned a large, level plot of land, previously known as "Scott's farm," north of the rocky Mesabi highland and about a mile from the old mill buildings. It was decided to locate the new community of Babbitt there. Pace Associates of Chicago were engaged as town planners, and Professor Robert Jones of the University of Minnesota acted as consulting architect. These men designed the new town of Babbitt, laid out the streets, business area, schools, and so on, and even submitted plans for some of the structures. Mr. Kingsbury, Reserve's vice-president, was in charge of the building program; the contract for the actual construction was let to John W. Galbreath and Company of Columbus, Ohio. Some of the old buildings were repaired and put into temporary use for the contractors' crews, but most of the workers lived in trailers near the site.

The cost of adequate houses in this climate was estimated at from $12,000 to $16,000 per unit, excluding service facilities and grading. Studies made by Pace Associates showed that the existing population within a radius of ten miles of Babbitt had a density of only 0.7 persons per square mile. Thus the firm concluded that, since only a very small percentage of the needed employees were already living in the area, most of them would have to commute from other range towns or would require housing at Babbitt. For the operation of the small mine and the Babbitt plant, about 175 employees would be needed. An increase to over 1,000 permanent, year-round jobs was expected by 1955 when the expanded project would go into production with the much larger mine, the coarse-crushing plant, and the railroad. After considerable discussion of all these factors, Reserve decided to build eighty houses at Babbitt for occupancy in 1952; more units would be added as they were needed.

The new townsite was fairly level and required very little grading. The topsoil was good and below it lay several feet of gravel and sand, making the installation of water and sewer pipes simple compared to the difficulties encountered in old Babbitt. Water was secured from deep wells drilled nearby, and complete water treatment and sewage disposal plants were built. All facilities were provided for an eventual town of 1,288

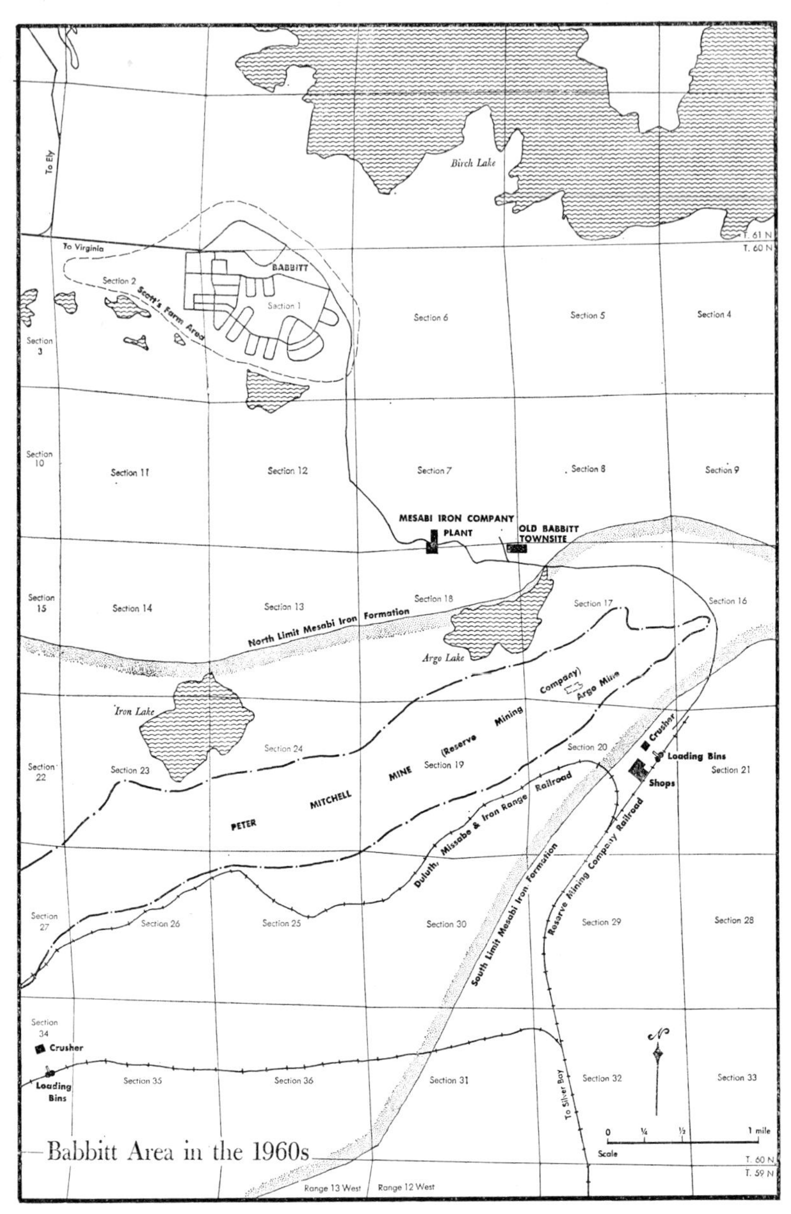

FIGURE 10. *The Babbitt area in the 1960s, showing the relationships between Reserve Mining Company's new town and mine and the Mesabi Iron Company's earlier town and installations.*

dwellings, although space was available to the west for twice that number. Babbitt and the other taconite towns are unusual in Minnesota in that they were completely planned and laid out before construction was started.[5]

It was interesting to watch the town develop. At first some houses were built on the site, and then prefabricated houses began to arrive weekly. They were built at Biwabik by Model Homes, Incorporated, and were moved to Babbitt on great double trailers with all conveniences installed, including plumbing, electric wiring, washers, and dryers. Some said they even came equipped with furniture, a girl, a wedding ring, and a license, if you wanted them. The foundations were already completed, and when the houses arrived, they were set in place. As soon as the pipes and electricity were connected, the new occupants moved in. It was not unusual to see a bulldozer grading around a foundation in the morning and by evening to find a house on the site with the lights glowing and evergreen branches in the window. The appearance of the evergreen boughs was in accordance with a local custom, apparently Scandinavian in origin, to bring good luck to a new venture.

In 1952, when the families from New York and Kentucky first arrived, Babbitt was not a very attractive place. A few new houses had been built, and as rapidly as others were completed, families moved in, but the streets were unpaved and there were no trees or sidewalks. When Mr. Kelley inspected the town, he immediately ordered all the yards sodded and the streets paved. After this was done, the village moved right up out of the mud and the children had a place to play.

Nevertheless, the first winter of 1952–53 was pretty tough. There was no entertainment except what the people themselves arranged — not even a movie closer than Ely. Going there or to Virginia became an exciting occasion. Fortunately about that time television transmission at the Duluth stations was improved, resulting in good reception at Babbitt. TV entertained the children part of the time, and kept the people from feeling quite so out of contact with the world. The isolation was worse for the single men and those whose families were still "back

East." Most of them liked their jobs, but after work there was nothing much to do except get together and play poker.

Ed Furness, the plant superintendent, was as lonely as the rest, but he had a lot of fun getting to know the reserved "Finlanders" from the Embarrass Valley. A number of people from neighboring communities got jobs at the plant. Men from Embarrass, Ely, and even from as far away as Virginia and Grand Rapids, drove to Babbitt singly or in car pools. One morning Ed came into the mill office all smiles and said, "Well, I have finally accomplished something I set out to do three months ago. Every day when I come to work Toimi is sweeping at the top of the stairs. When I first started speaking to him, he wouldn't even look up. At the end of about a month he did look up once. A while back he grunted at me, but today was the crowning achievement. When I said 'Good morning, Toimi,' he said 'Good morning' back to me."

Ed also recalls one of the Kentucky men looking out a window in the pelletizing plant during that first winter. He had rubbed some of the dust off the glass and was watching the first big snowstorm of the year. Thinking he needed cheering up, Ed walked over to him and asked, "Beautiful, isn't it?"

"Yes, I guess so," Jimmie answered. "Damned cold though."

"Oh, it isn't so bad when you get used to it," Ed told him encouragingly.

"Well," said Jimmie doubtfully, "maybe not, but it must take a long time to get used to it. The Finlander girls who've been here all their lives wear an awful lot of clothes even at night."

Reserve rented the new houses at something like $65.00 a month, but the occupant could buy them for $10.00 less than that per month with no down payment. Eventually almost every family owned or was buying its home. The proud owners planted flowers and trees, and Babbitt rapidly became an attractive town with a shopping center, schools, and churches. The company built a one-story grade school, which opened on September 8, 1954, with a gymnasium, a library, and hot lunch facilities.[6]

At first, Frank Emanuelson, who had been the Mesabi company's manager at Babbitt all these years, served as town supervisor for Reserve, but after a village was organized and incorporated on September 12, 1956, Reserve turned all the facilities over to the new officers. Adam Karakash was elected mayor, and Frank Petric, Harold Wahlstrom, Daniel J. Clemmer, and Frank L. Marinaro were the first councilmen. Mr. Emanuelson served as chairman of the school board. Before long, the school constructed by the company had to be enlarged, and in October, 1960, a second one was dedicated which also included a high school. The town was now complete, self-supporting, and "on its own." [7]

As churches, clubs, and baseball teams were organized, the people began to feel more at home.[8] The first summer many of the out-of-state families could hardly wait for their vacations so they could visit their old homes. But soon they came drifting back to Babbitt — sometimes before their vacations were over — one man to play on his baseball team against Ely, another to look after his garden, and all of them with a fresh viewpoint on the comparative simplicity of their former jobs. One man returned with a black eye. When he was asked how he got it, he said, "Those fellows back there in that little bit of a plant bragged about relining a crusher every two months, and when I told them we relined one every two weeks, they wouldn't believe me. I had to convince them."

In the Babbitt test plant, rock hauled from the mine was dumped into the primary crusher, a powerful machine driven by a 300-horsepower motor, which simply squeezed three-foot blocks of taconite until they crumbled into nine- to twelve-inch pieces. From this crusher, the taconite dropped into a second one which broke the rock into three-inch pieces. At this size the rock was conveyed to two additional crushers in which it was reduced to about three-quarters of an inch in size. It then went to storage bins from which it was fed to the rod mill.

The rod mill originally installed at Babbitt was ten and one-half feet in diameter and twelve feet long. It was charged with about seventy tons of round steel rods. From the rod mill, the product went as a thin mud to the magnetic cob-

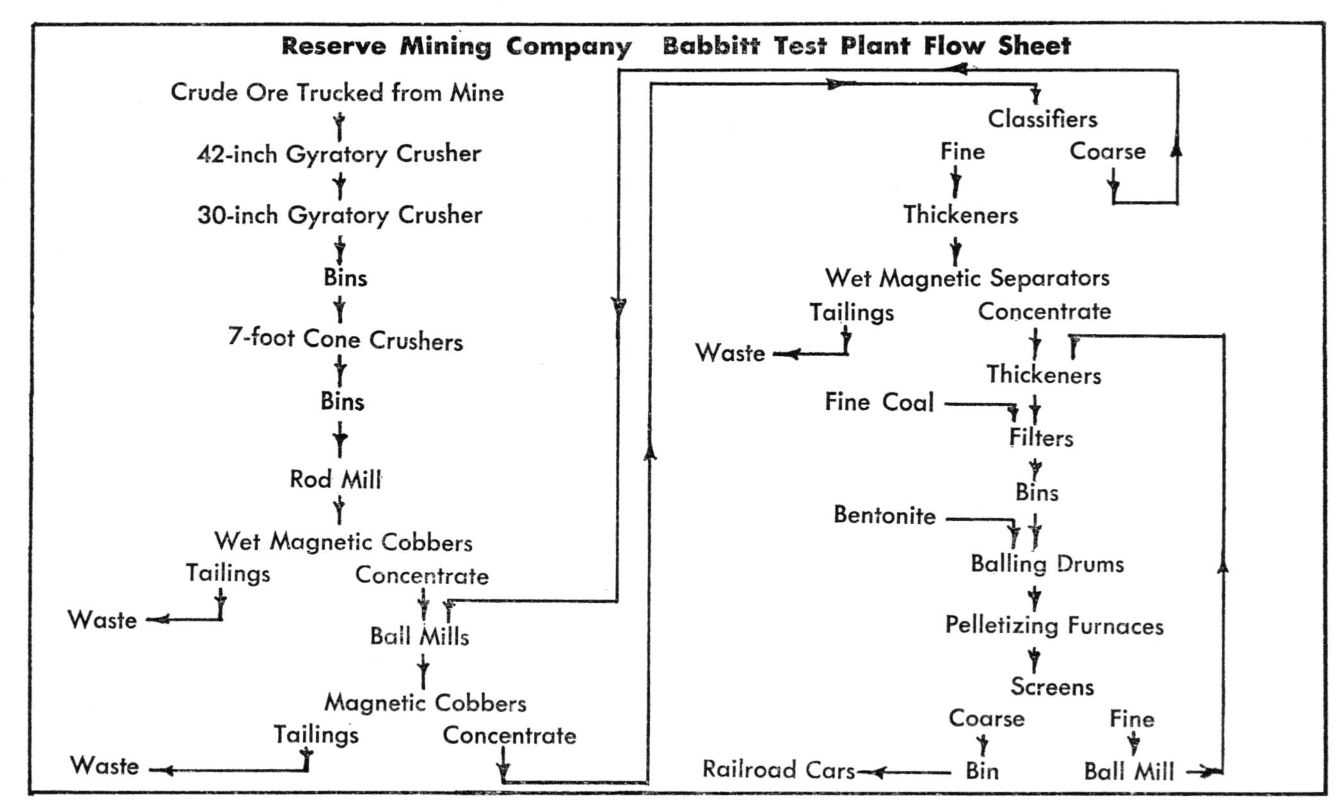

FIGURE 11. *The flow sheet used in Reserve Mining Company's Babbitt test plant in the 1950s.*

bers where the nonmagnetic particles were rejected, reducing the bulk by about one-third. The higher-grade portion of the ore then went to the ball mills. In the Babbitt plant, these mills were charged with about eighty-five tons of steel balls having a maximum diameter of two inches.

The wear on both the rod and ball mills was severe, and the kind of steel needed came in for much study and analysis. The lining plates, balls, and rods must be hard enough to resist wear but not brittle enough to break. A manganese steel alloy was found to be satisfactory; it is both hard and tough, but not so tough that the rods will bend, for bent rods will not grind and will cause other difficulties in the mill. The steel wear in the rod mill was about one pound per ton of taconite or 3,000 pounds daily. New four-inch rods weigh approximately 500 pounds each and about six of these were charged daily. In the ball mills the steel wear was a little over half a pound per ton of taconite or about 1,700 pounds daily. New two-inch balls were added with the ore without shutting down the mill. Each four-inch rod stayed in the mill about five months before it wore out, and during that time it crushed enough taconite to produce 150 tons of concentrate. Similarly, each two-inch ball remained in the mill about six months. During that time, it ground enough taconite to make about one ton of concentrate. Optimum conditions for both types of mills were important, because of the high cost of their operation.[9]

The Babbitt plant was designed to produce about 40 tons of filtered concentrate per hour, assaying 64 to 65 per cent iron, from 120 tons per hour of original rod mill feed, assaying about 24 per cent iron, but trouble developed in the grinding operations soon after the test plant started up. The rod mills did not grind the required amount of feed, and the rods in them seemed to be tossing about haphazardly rather than rolling in an orderly way. Much advice on how to correct this was received from outside experts. One of them said that the rods were too large. He advised speeding up the mill and using rods no larger than two inches in diameter. To try this, the mill was shut down, and seventy tons of rods were removed and sorted — a task that took a crew of men one whole night and the following day. The mill was started again early the next evening with small rods and at a slightly increased speed. By the time the expert returned the following morning, the mill was completely filled with coarse, unground taconite rock and a mass of bent and tangled rods, which the men in the plant call a "bird's nest." It took nearly a week of hard, dangerous work to get the mess untangled and the rod mill cleaned out. Ed Furness was so disgusted he said if they ever let another "expert" into the plant, he was leaving for home, which to him was Lyon Mountain, New York.[10]

This was the end of the imported experts who were consulted on the design and operation of the Babbitt plant. Doc Kelley explained that their assistance would no longer be needed. Then he turned the plant and its problems over to Bob Linney and Ed Furness. They immediately slowed down the speed of the mill and loaded it with four-inch rods, exactly the reverse of the expert's recommendations. The mill ran beautifully. But at the desired feed rate, the amount of coarse oversized rock particles was still somewhat greater than desired, and the rods still had a tendency to bounce around in the mill instead of rolling properly.

After weeks of study and experimentation, Ed and Bob decided that the mill was too short for its diameter. Consequently, a section 2 feet, 7 inches, in length was added in 1953. This was an expensive change, but it solved the problem. The plant immediately went into smooth operation at the desired capacity of 120 tons per hour.

This experience illustrates the advantages of operating a pilot plant with full-sized commercial equipment before a final large plant is built. When we suspected that the rod mill at Babbitt was too short, it was a comparatively simple matter to shut down the plant and, as an experiment, increase the length of one mill. In a large plant, however, having a dozen or more such machines, the installation of short, high-speed mills would have been a disaster.

Similar information regarding the sizes and types of various pieces of equipment in the Babbitt plant led to modifications in the design of other machines for the big plant at the lake, but probably the most important information to come out of the Babbitt plant pertained to pelletizing. While I was on leave from the Mines Experiment Station, I was more or less responsible for the design of the pelletizing section of the Babbitt plant, assisted by engineers of the Bros Boiler and Manufacturing Company of Minneapolis and the Swindell-Dressler Corporation of Pittsburgh. We were instructed to set up an installation that would provide a thousand tons of pellets per day, using the best equipment developed up to that time. A large area of the plant was to be left vacant to test new furnace designs, but at the moment we were not to concern ourselves with such experiments. As a result, we used as a model the furnace developed by Mr. Steffensen and his staff at Bethlehem, for it was, in 1951, the only one operating satisfactorily.[11]

We made the Babbitt furnaces 37 feet high, the same size as those at Bethlehem, but we omitted the center partition, giving us a single rectangular shaft 5 feet wide by 10 feet long. We erected four such furnaces, each of which was expected to develop a capacity of 250 tons of pellets per day. The furnaces were equipped with chunk breakers, discharge feeders, and two combustion chambers designed for heat recuperation. They burned oil.

We also installed four balling drums, each of which was seven feet in diameter and equipped with a reciprocating leveling bar. To carry the green balls from the drums to the furnaces, we built feeders which were of rather elaborate design and worked on the pantograph principle. They were about the only new and completely untried piece of equipment in the Babbitt plant. As it turned out, they operated well enough, but they took up a great deal of space.

Babbitt had the first multiple-unit pelletizing installation ever put together. We made a number of mistakes in laying it out, and it was an unhandy, dusty plant, but it did make pellets, eventually at the desired rate of a thousand tons a day. The first pellets were produced on August 28, 1952, and within a few days they were being shipped by rail to Republic's plant at South Chicago.[12]

Space had been left at Babbitt to experiment with some type

of pelletizing furnace that would be suitable for the large lakeside plant. The shaft furnaces we installed at Babbitt had such a small capacity that 135 of them would have been required for a 10,000,000-ton plant at the lake. Equipment five or ten times larger was needed, but no such pelletizing furnace had ever been built.

The Mines Experiment Station had been testing updraft burning of the balls of concentrate on a traveling grate machine, and Allis-Chalmers had continued to study downdraft firing. Fred Darner and a group of Republic and Armco engineers kept track of both sets of experiments. With engineers of the McKee company, they then designed a traveling grate machine which combined updraft and downdraft burning.[13] A large machine of this type, having an expected capacity of a thousand tons a day, was built and installed at Babbitt. It was equipped for updraft drying at the feed end, followed by downdraft ignition and burning. Generally, it worked well, made good pellets, had the desired capacity, and was not difficult to operate. The lakeside pelletizing plant was designed to use these new machines.

By the end of the first year, the Babbitt plant was operating so well that we invited the members of the Minerals Beneficiation Division of the American Institute of Mining and Metallurgical Engineers to visit it on August 31, 1953. Several hundred of them came, including a group from Erie. We took them through the plant and showed them everything. This visit broke down the "iron curtain" that Erie had put up around itself. After that, contact between Reserve and Erie technicians gradually became freer until scarcely a week passed that some men from the two plants did not visit back and forth. Ever since that time the exchange of information among Reserve, Erie, and Oliver has been free and cordial.

Although the Babbitt plant cost Reserve several million dollars, it supplied information which amply repaid its cost. In addition, men trained in the test plant and mine at Babbitt were available for key jobs when the big installation at the lake went into production in 1955. After that, having served its purpose, the test plant was closed on October 25, 1957.[14]

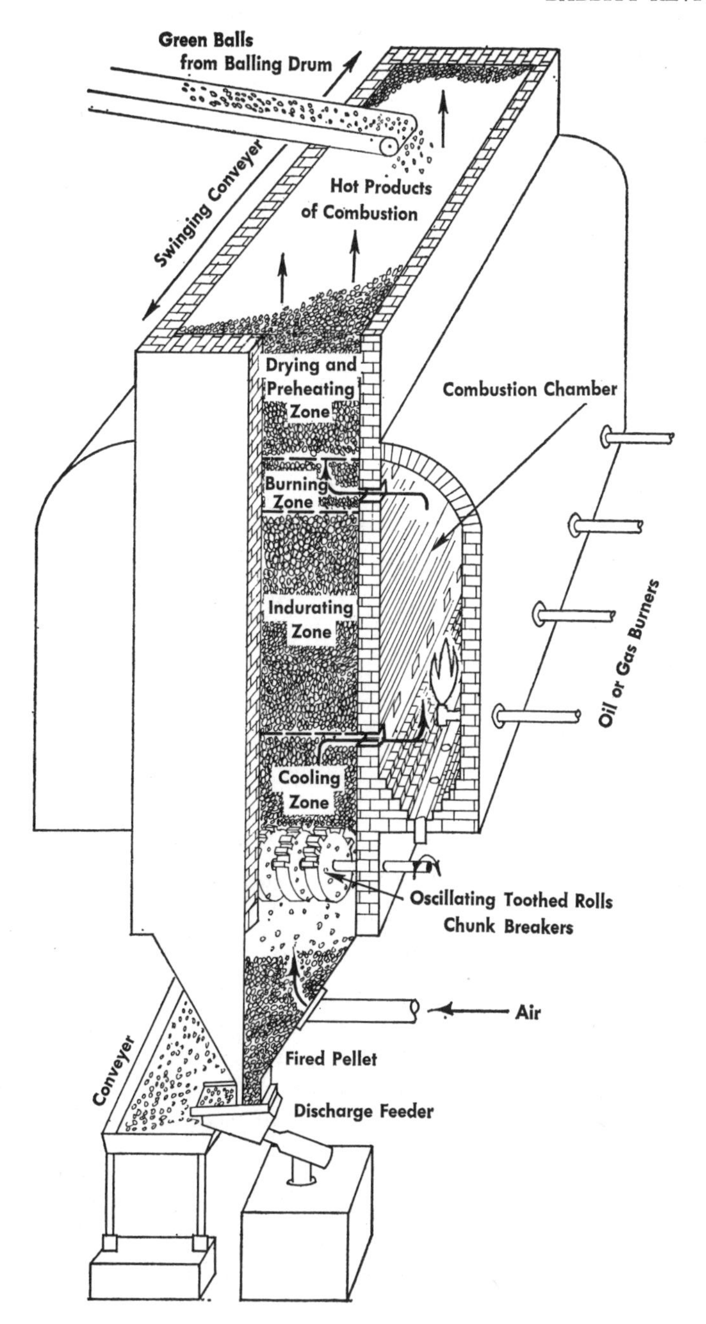

FIGURE 12. *The pelletizing furnace in the Babbitt test plant.*

10 Silver Bay Is Born

While the pilot plant and new town were being built at Babbitt, work was also progressing on the railroad and on the big lakeside plant and harbor. A construction contract had been let by Reserve in 1951 to a firm that came to be known as "HAD." It was composed of the Hunkin-Conkey Construction Company of Cleveland, the Arundel Corporation of Baltimore, Maryland, and the L. E. Dixon Company of San Gabriel, California — three experienced companies brought together by S. E. Hunkin to construct this $185,000,000 project. A supervisory staff assembled from the three firms made its appearance on the north shore of Lake Superior, living for a time at Two Harbors and in some of the old houses on the plant site near Beaver Bay. Edwin C. Lampman, who was Reserve's construction manager, co-ordinated and expedited the work of the contractors and subcontractors at the lakeside and Babbitt plants and on the railroad connecting the two.[1]

As time went by several thousand construction workers arrived. They were housed principally in a large dormitory and in a trailer camp near the plant site, and they were fed in a mess hall presided over by the chief of the commissary, big, roly-poly "Pat" Unger. Similar but smaller establishments were set up at Babbitt and at Jordan — about mid-point on the railroad — where there were also camps for the small army of workmen.

THE NEW TOWN OF BABBITT *was built in northern Minnesota by Reserve to house employees of its test plant and nearby Peter Mitchell Mine. The first families arrived in 1952, and the village was incorporated in 1956.*

THE OLD BABBITT PLANT *of the Mesabi Iron Company was enlarged and rebuilt for use as a test plant by Reserve Mining Company in 1951–52. It operated from 1952 to 1957 producing taconite pellets.*

AN AERIAL VIEW *of Reserve's Peter Mitchell Taconite Mine as it looked in 1963.*

A ROCK FACE *in the Peter Mitchell Mine in 1963 shows the large pieces of taconite that have been broken by blasting.*

A JET-PIERCING MACHINE *sinks blastholes in the hard taconite by using kerosene mixed with oxygen to generate a jet flame at a temperature of 4300 degrees Fahrenheit.*

WIRING A BLASTHOLE *in the Peter Mitchell Mine to blast taconite rock down into the mine pit.*

DETONATING A BLAST *at the Peter Mitchell Mine. On August 25, 1958, what is said to be the largest open pit blast to date was set off at the mine when 794 holes were exploded breaking loose 1,140,135 tons of taconite.*

LARGE ELECTRIC POWER SHOVELS *lift five cubic yards of taconite at a time and load it into forty-five-ton, side-dump trucks to be hauled from the mine to Reserve's coarse-crushing plants at Babbitt.*

TRUCKS *dump the taconite into the large primary crusher at Babbitt. This huge crusher, which is set 167 feet deep in solid rock, reduces the taconite to pieces approximately nine to twelve inches in size. After further crushing, it is taken by rail to the E. W. Davis Works at Silver Bay.*

THE SITE *of Reserve's E. W. Davis Works at Silver Bay as it looked before construction began in the winter of 1951–52. Beaver Island is at the right.*

THE E. W. DAVIS WORKS *in 1959 before the plant was expanded. The new town of Silver Bay, which was built by Reserve, may be seen in the background.*

THE FIRST TRAINLOAD *of coarse taconite from Babbitt reached Silver Bay in August, 1955.*

THE AUTHOR *pulled the switch to dump the first carload of taconite into the big plant at Silver Bay.*

DUMPING *the first car at Silver Bay. This huge rotary car dumper picks up the loaded railroad car with its 87 tons of taconite and empties it without uncoupling the car. The ore then goes by conveyer to the fine-crushing plant. Note the evergreens.*

IN THE FINE-CRUSHING PLANT *the tumbling and rolling rods in the rod mill reduce the taconite to the consistency of fine sand. The end of this mill has been removed.*

A TUNNEL *under Highway 61 conveys the ore to the rod mills.*

THE PICTURE *at the far right shows the mills which grind the taconite rock to the fineness of flour. In the center photograph are magnetic separators and hydroseparators which remove the particles of magnetite and discard the silica. The finisher magnetic separators (below) complete the concentration process, and filters (at left) remove the water from the fine concentrate.*

FROM THE FILTERS, *the concentrate is conveyed to the balling drums and formed into pellets less than a half inch in diameter.*

FROM *the balling drum, the pellets go to furnaces where they are fired at 2300 degrees Fahrenheit to harden them and change the magnetite to hematite. At left, pellets are shown emerging from a furnace at Silver Bay.*

AN ORE BRIDGE *carries the finished pellets to the stock pile. In winter the pellets are stored for shipment down the lakes with the reopening of navigation in the spring.*

PELLETS *of many sizes and shapes can be produced in present-day taconite plants to meet the specifications of blast furnace operators. Those shown vary from two inches in diameter (lower right) to one-eighth inch (lower left).*

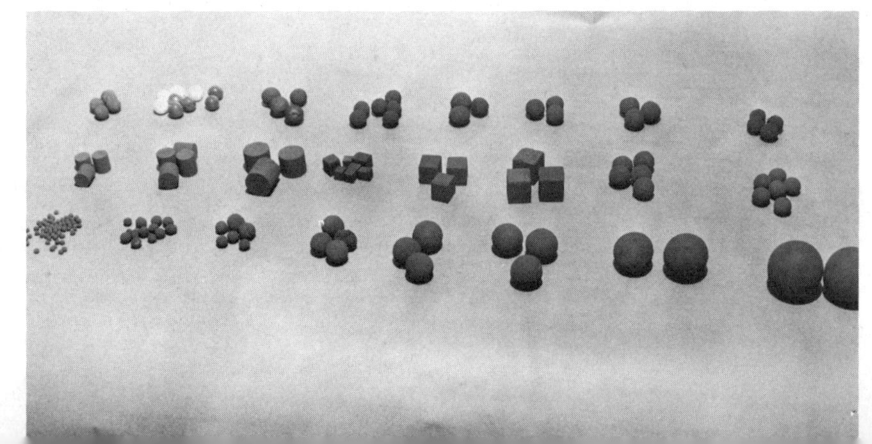

THE "C. L. AUSTIN" *carried the first cargo of 10,800 tons of pellets down the Great Lakes from the E. W. Davis Works at Silver Bay in April, 1956. Belt-type loaders pour the pellets from storage bins directly into the holds of waiting ore boats for shipment to Eastern blast furnaces.*

THE TEST *which conclusively established the superior smelting qualities of taconite pellets was made in 1960 in the Bellefonte Furnace (left) of Armco Steel Corporation at Middletown, Ohio. At right is the even larger Amanda Furnace which was built to utilize taconite pellets. Kenneth R. Haley (left below) was in charge of the 1960 test, and James L. Brady (right) conducted the first smelting test of taconite pellets in the Mines Experiment Station's small blast furnace in 1948. Courtesy Armco Steel Corporation.*

The big push in 1951–52 was to get the railroad built between the plant site at the lake and Jordan—where it would cross the tracks of the Duluth, Missabe, and Iron Range Railway from Two Harbors—in order to bring in the equipment and construction materials needed. The right of way had been relocated several times and had been checked by engineers of the Baltimore and Ohio Railroad. Several rock cuts were necessary and some muskegs had to be dug out and backfilled with rock and gravel. The tracks at the lake would start at an elevation of 600 feet and climb 47 miles to Babbitt at an elevation of about 1,500 feet. The taconite would be hauled downhill, the steepest part of the slope being near the lake shore. Ultimately, the railroad would have to haul to the lake 30,000,000 tons of taconite per year, or an average of over 80,000 tons per day. This would be a thousand carloads, or ten big trains a day, between Babbitt and the lakeside plant. (By the summer of 1964 as much as 90,000 tons a day was being hauled.)

While Oglebay Norton and Gus Wynne had originated the general plan for the whole Reserve project, detailed designs for the lakeside plant were being prepared in Cleveland. About the time the Babbitt test plant went into operation, Mr. Kelley began to hold weekly meetings of the whole staff at Babbitt. He flew in from Cleveland in the company plane to preside at them, usually bringing with him Fred Darner and others from the Cleveland office. Some of the Oglebay Norton men from Duluth attended, and Oscar Lee and I were usually there.

For a time it was not quite clear whether the big, new lakeside plant was being engineered under the direction of Gus Wynne, Fred Darner, Oscar Lee, Bob Linney, or Henry Martin. As soon as Doc Kelley started his weekly meetings, however, there was no longer any question of authority. He whipped the group of prima donnas into an efficiently functioning team in a way I never dreamed possible. At the meetings we often wrangled over a point that Bob Linney wanted one way and Fred another, with Ken Haley, Oscar Lee, and I finding fault with all of them and each other. Then Doc would say, "All right, you fellows can't agree, so I will have to decide. Anyone who is not willing to give wholehearted support to my decision, whether he likes it or not, just get up and go back home now. I don't want you around." Then he would announce his decision, and the point would be settled once and for all.

Fred Darner was given the job of engineering and designing the various units and seeing that they were built according to the plans and within the estimated cost. Fred was well acquainted with the problems of designing steel plants, but since this was the first taconite plant to be built, he had no experience with such installations. He was an energetic and perceptive person, and he visited many concentration plants, studied mine layouts, and quickly acquired all the information he needed about ore processing in general and taconite in particular. It was disconcerting to explain to Fred the operation of some new piece of equipment we had heard was being used in some remote plant and have him wait politely until we were all finished to tell us that he knew about it, that he had seen it, and that he already had a drawing of it. He was a meticulous worker and his drawings and cost estimates were models of detail and accuracy, but he was not an easy man for the operating staff to get along with. His objective was to build the best possible plant to do a definite job at minimum cost; the operating staff was interested in flexibility and ease of operation. As a result, there was a constant battle between Fred and Bob Linney, even though the two men had the highest respect for each other. Doc Kelley was the final arbiter, and his decision in the wrangling was respected as final.

Clearing of the site on the shore of Lake Superior began in 1951. Bulldozers could be used in some places, but in others, where the slopes were rocky and steep, the work had to be done by hand. A great deal of rock blasting was necessary not only for the foundations of the plant but also to relocate Highway 61, which cuts through the site. The blasted rock was used to build breakwaters from the shore to two small islands, which were conveniently located to effectively enclose a harbor that would protect shipping from the heavy storms that sweep down Lake Superior from the northeast. At the request of the Coast Guard, the smaller of the two was officially named Pellet Island in 1955.[2] It was so small, steep, and rocky that it could be used only for a navigation light, but the second one, known as Beaver or Pancake Island, was larger. An early owner had built a small, wooden shack on it and reportedly spent his honeymoon there.

At first, plans called for the blasting of some rock on Beaver Island for use in building the breakwaters. When this became known, Reserve received a letter from Dr. Olga Lakela, a professor at the Duluth Branch of the University of Minnesota, asking that Beaver Island be preserved. In May, 1948, she had written an article, entitled "Ferns and Flowering Plants of Beaver Island, Lake Superior, Minnesota," for the *Bulletin* of the Torrey Botanical Club. In her letter she said that she had visited the island many times collecting specimens of the rare plants and ferns that grew there. One in particular, the closed bluebell, she wrote, was "new to science" and the island was "the only place where it is known to grow." Consequently, Reserve changed its plans in order to leave the island intact. At last report, the closed bluebells, the ferns, and a herring gull colony were all doing nicely.[3]

Construction of the harbor followed in general the recommendations of the United States Army Corps of Engineers. A large-scale model had been made for Reserve by the Waterways Experiment Station at Vicksburg, Mississippi. A small amount of underwater drilling, blasting, and dredging was necessary in order to provide a minimum water depth of thirty feet. The breakwaters, which were made wide enough to permit a maintenance road along the top of them, rose about fourteen feet above the surface of the water.[4]

The harbor, which was more than half completed by the fall of 1952, was built exclusively for the purpose of shipping pellets and receiving coal, oil, and other supplies needed by Reserve. Although it was to be a private harbor, it was approved by the United States Coast Guard as a refuge for any craft in trouble. Compared to others along the north shore, Reserve's harbor seemed small. No space was provided for anchorage, and there was room for only two or three ships to tie up at the quay for loading or unloading. Every facility was provided, however, to rapidly load the pellets into the ore boats. The harbor was designed to load an average of 60,000 tons of pellets daily during the shipping season, which would mean an average of four or five boats a day.

The weather during the winter of 1951–52 was favorable for construction. By spring some twenty miles of railroad right of way were cleared, and the relocation of the highway along the north shore was virtually completed.[5] Excavation for the plant buildings went ahead day and night during the summer of 1952. Everything was laid out on the basis of 10,000,000 tons of pellets per year, although only one-third of the plant was to be constructed at this time. Generally, blasting and actual rock removal were carried on in anticipation of future expansion. By late 1952, sand and cement were being brought in on the new railroad, and foundations for the buildings and equipment as well as the new town for employees were beginning to take shape.

Like Babbitt, the still unnamed town at the lake had been designed and laid out by Pace Associates. The total number of employees was expected to be slightly higher than at Babbitt, but the population density along the lake shore was also greater. Thus the planners concluded that a good number of workers would commute from Two Harbors on the west and Grand Marais on the east. A town of 1,500 families was planned, but it was decided to build a smaller number of houses until the need could be more accurately determined.[6]

The town was laid out in a large, natural amphitheater lying between high, rocky hills paralleling the Lake Superior shore. The village would be rather long from east to west but narrower from north to south. The amphitheater in which it was located connected with a second parallel valley to the north that could be developed for future expansion if needed. The soil — what there was of it — was a dense red clay. In some places, it was many feet deep, and in others there was no topsoil — just solid, bare rock. This necessitated considerable grading and some blasting in order to install water and drain pipes and build house foundations and roads.

The clearing and rough grading of the townsite was done by

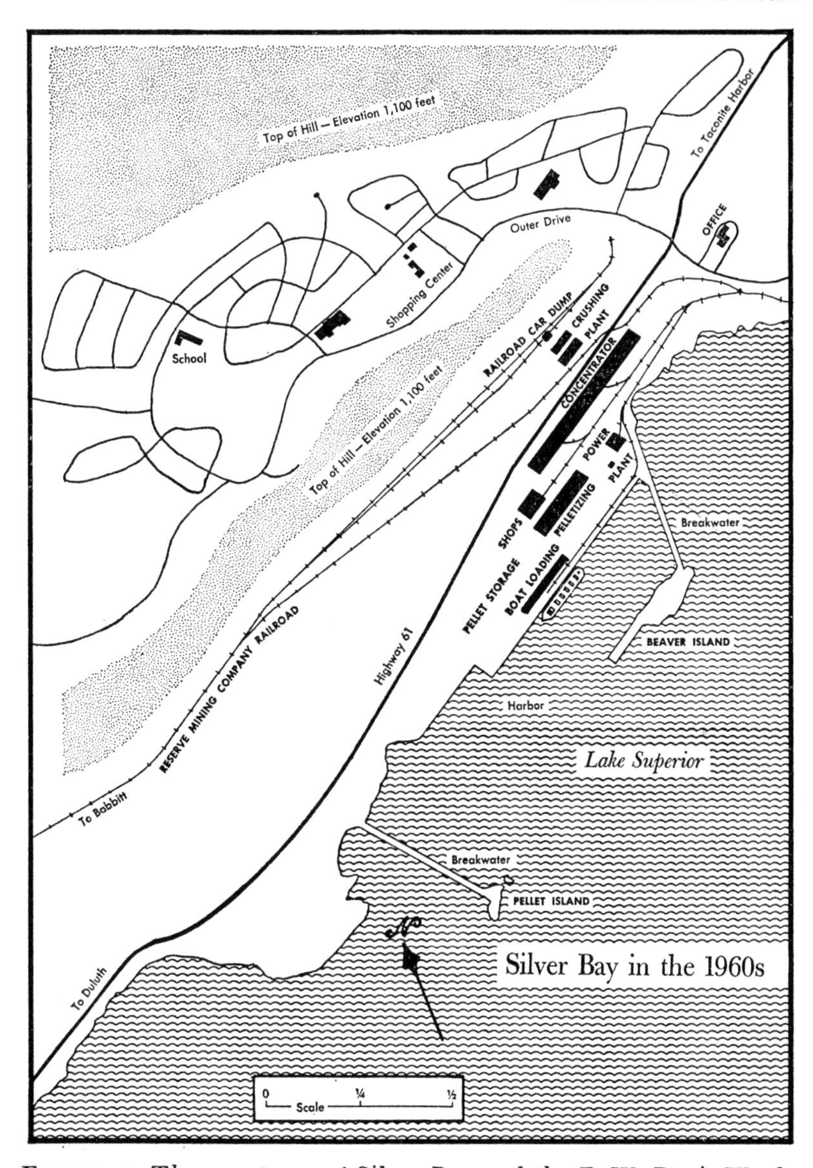

FIGURE 13. *The new town of Silver Bay and the E. W. Davis Works of Reserve Mining Company in the 1960s.*

HAD with its powerful bulldozers and trucks, but later John W. Galbreath and Company completed the town and supervised the erection of the buildings. As at Babbitt, this organization later took over the management of the town and the rental and sale of the houses to Reserve employees, the program being about the same as that at Babbitt with about the same result—practically every employee with a family decided to buy his house. Two apartment buildings were erected almost at once to accommodate the men who expected their families to join them later, and a few residents moved in as early as December, 1952.

Completely built houses could not be brought to the new town as they were at Babbitt because of the narrow bridges along the route, but great trucks brought in many houses in panel form. These were rapidly assembled on the foundations prepared for them. Several models were available with or without basements, with one, one and a half, or two stories, or with split levels.

The employees could choose the type of house they wished to buy, without a down payment. Reserve paid the cost of installing utilities, landscaping, and paving, and even absorbed 10 per cent of the cost of the house itself. The company also erected a shopping center and a large grade school. Later a second grade school was built, and in November, 1958, a high school was constructed and named for Doc Kelley.[7] Nothing could have been more fitting. Outside of the plant itself, Doc's greatest interest was the children of the community. He often said: "There are two things that I cannot forgive a man—neglect of his job or neglect of his children."

On May 1, 1954, it was announced that Reserve's new town on Lake Superior was to be called "Silver Bay." This had been the name of the bay in which Reserve's harbor was being built and the name of a store and resort formerly operated there by Oscar Pedersen and later by Joe Betzler. One of the first houses to be completed was assigned to Mrs. Arnold Erickson who conducted Silver Bay's first post office in her home. On May 1, 1954, she received a stamp for canceling the mail marked "Silver Bay." For a few days she did a big business selling post cards and covers and canceling them with the new stamp still dated May 1.[8]

The pattern of settlement at Silver Bay was unlike that of Babbitt for the simple reason that many of the first families to move in came from Babbitt. Men who had learned to operate and maintain the machines in the Babbitt pilot plant were transferred to Silver Bay. For this reason, a group of people who knew one another and who had experience in forming a new community moved to Silver Bay at the beginning. Other employees more or less looked to the Babbitt "old-timers" for direction and followed their lead in organizing the new town. Moreover, Silver Bay was not isolated. It was on a well-traveled highway leading to the Minnesota canoe country and Canada, and many tourists passed that way, especially in summer. As a result, Silver Bay never had as much of the frontier atmosphere about it as Babbitt did.

On October 16, 1956, the village of Silver Bay voted to incorporate. The first election was held on November 27, 1956. Len Brickles became mayor and Theodore Matthews, Roy Jacobson, Peter V. Bromme, and Walter H. Frey were elected trustees.[9]

As at Babbitt, the first church to be erected was the Lutheran. It was followed by the United Protestant (composed of over a dozen denominations) and the Catholic. Clubs were formed, as were such civic and fraternal organizations as the Rotary, the Junior Chamber of Commerce, the Lions, the Veterans of Foreign Wars, the Masons, and others.[10]

When the Masons held the first regular meeting of Taconite Lodge No. 342 on March 10, 1957, the officers had received some furniture as gifts from other Masonic groups in Minnesota. One of the important symbols of the lodge consists of two pieces of stone — one rough and one polished. Those acquired by the new Silver Bay organization were of limestone. It seemed to me that a Taconite Lodge should have stones of taconite, but this proved to be easier said than done. Reserve donated pieces of taconite from the mine at Babbitt, and we sent them to the Cold Spring Granite Company at Cold Spring, Minnesota, for cutting and polishing. The firm quickly reported, however, that the pieces were full of small cracks, apparently the result of blasting the rock in the mine, and that these made cutting and shaping impossible. The company asked that we send some pieces of rock that had never been blasted. Floyd Erickson, who was in charge of Reserve's mine, was a good Mason. He laboriously pried some pieces of taconite from a rock outcrop that had never been blasted and sent them to Cold Spring. These were cut and worked into the desired shapes without serious difficulty, although the company found them considerably harder than the granite it normally used. Taconite polishes beautifully, showing plainly the alternate richer and poorer bands of minerals.

About this time members of the new St. Mary's Catholic Church in Silver Bay decided that its altar and baptismal font should be made of taconite. Consequently when Reserve received a second request for two large blocks of taconite to be shipped to the Cold Spring Granite Company, Mr. Erickson knew that the pieces should be secured from an outcrop that had never been blasted. From those he sent, a beautiful and unique altar was made. The top slab, weighing about five thousand pounds, is some three feet wide by six feet long and highly polished. The baptismal font is truly a work of art, combining both rough and polished surfaces.[11]

The north shore of Lake Superior was a busy place during the summers of 1952, 1953, and 1954 while both Silver Bay and Erie's Taconite Harbor were under construction. Observation and parking areas were set up at both locations, and tourists by the thousands stopped to view the work, get descriptive folders, and ask questions of the attendants.

In July, 1953, the directors of Reserve met at Silver Bay and announced that the lakeside plant was to be named the "E. W. Davis Works," in recognition of the assistance given to the project by the author and the Mines Experiment Station. They presented me and the university with beautiful hand-illuminated books containing testimonials of appreciation. I still treasure my copy, and the one given to President Morrill is now in the university archives at Minneapolis. So far as I know, this is one of the few times on record that a large corporation has in this fashion expressed its appreciation to a university research organization for its contributions to the success of a commercial enterprise.[12]

By this time I was, of course, back at my post at the Mines Experiment Station, although I made frequent visits to the north shore to watch our years of research being transformed into an operating unit. After I retired in June, 1955, I moved to Silver Bay to become a consultant for Reserve. Thus with a group of other local residents I was on hand to see the first trainload of taconite reach Silver Bay from Babbitt on August 29, 1955. Its arrival was slightly delayed because Kenneth Kurry and his helpers at Babbitt had taken time at my request to cut small evergreen trees and place them on the engines and in the cars. Doc Kelley and Harold D. Barber, superintendent of the railroad, were impatient when the train did not arrive on schedule, until I explained that in October, 1892, the first trainload of iron ore shipped from the Mesabi Range arrived at Superior decorated with evergreen trees. After Mr. Kelley had stormed about the foolish delay, he radioed the engineers to keep the whistles blowing, so the train came down the track with a great lot of noise and with small evergreens standing upright in the ore cars and tied to the fronts of the two big diesel locomotives. I was allowed to operate the controls, under the careful supervision of Ed Furness, and dump the first car, No. 132, of eighty tons of Babbitt taconite into the new plant.[13]

After that, the world's first large commercial taconite plant began to hum, even though parts of the concentrating and some of the ore-loading facilities remained to be completed, and none of the pelletizing machines was quite finished. Until they began to operate, the concentrate was hauled away in trucks and stock-piled.

The start-up period of such a new and complex operation is very critical. This was especially true for Reserve. While its plant was being completed by members of the Building and Construction Trade Unions of the AFL (American Federation of Labor), operations were being started by the operating personnel, who were members of the United Steelworkers of America (CIO). Such circumstances required careful planning of the work to be performed by each group in order to avoid jurisdictional disputes between the unions which could easily have stopped all progress.

Every job had to be carefully described and classified in accordance with the job evaluation agreement between the company and the union. Since taconite processing was new, there were few precedents to be used as guides. W. L. (Larry) Edwards, who was director of industrial relations, and O. H. (Mike) Tourje, who was chief industrial engineer for Reserve, had the big task of negotiating the job classifications with representatives of the union. A rigid division was made between operating and maintenance personnel. The former did not undertake major repairs or alterations; they ran the equipment and adjusted the controls. Each operating department—mining, transportation, concentrating, and pelletizing—had its own superintendent. If the equipment did not operate properly, maintenance was to be notified. It was a large and well-organized separate department, under its own superintendent, consisting of such divisions as carpenters, mechanics, machinists, electricians, and welders.

All operations in the plant were planned on a twenty-four-hour-a-day basis seven days a week. Each man was scheduled to work a forty-hour week, and the various superintendents were held strictly accountable for proper scheduling to take care of emergencies that required overtime work. The average employee's pay was slightly more than $3.00 an hour with time and a half for overtime above forty hours a week.

The whole operation at Silver Bay and Babbitt required over 2,500 men and getting this organization put together and teaching the operators their jobs was no small undertaking. A little inattention could cause damage to equipment costing thousands of dollars. The initial plant at Silver Bay contained 192 magnetic separators and 36 rod and ball mills. In many instances expensive equipment was put into the hands of operators who had never seen anything like it before.[14]

The first pelletizing unit went into production in October, 1955. Some of the workers were local men who had previously made a living of sorts along the lake shore as commercial fishermen. They did not, of course, have the slightest idea what a pellet was or how a pelletizing machine worked. When the first pellets were discharged from the end of the big machine, one of these men was overheard asking a more experienced operator, "What's them things?" The answer was, "Them's what pays your groceries."

There were six parallel lines in the pelletizing plant—each consisting of three bins and feeders and three balling drums to feed one pelletizing furnace. As concentrate was available, the six lines were put into operation, one after the other. The last of the six started up in February, 1956.[15]

During the winter of 1955–56 all the pellets were put on the stock pile at Silver Bay, but with the opening of navigation in the spring, the first boat—the "C. L. Austin"—arrived on the morning of April 6, 1956, to carry some of them down the lakes to Republic's plant at South Chicago.[16] Captain Peter J. Peterson brought the Wilson Transit Company's ship into the harbor in a 45-mile-an-hour wind and snowstorm. Loading began at once but was later suspended until April 8, when the "Austin" left with a cargo of 10,800 tons—the first shipment of taconite pellets from the E. W. Davis Works. To celebrate the occasion Doc Kelley and I presented Captain Peterson with a new hat—a hard safety helmet like that used by all Reserve employees—with the name "Reserve" painted on it. This was a slight modification of the old Great Lakes custom of crowning the captain of the first vessel to come into port in the spring with a new hat and conferring on him the title of "Duke."

To be sure, the first pellets were not as good as desired, but day by day they grew better. Ken Haley and Don Cooksey knew from experience the requirements for making good pellets, and they went to work on them. To begin with the pellets came off the machines in a thin trickle, but after all six pelletizing machines got going the operators soon exhausted the stock pile and then began to call for more ore from the concentrator.

In the mine, the crushing plant, and in the concentrator, capacities were gradually built up to 33,000 tons of taconite a day, the tonnage rate for which the plant was designed. Although about 11,000 tons of concentrate were being produced daily, this was not enough to keep the pelletizing plant operating at full capacity. The tonnage of rock from the mine was therefore increased, with the result that trouble developed at several places in the flow sheet. Conveyer belts were overloaded and had to be speeded up. Some chutes had to be rebuilt to handle greater quantities of ore and water. Feeders were speeded up, and other minor changes were made to accommodate greater tonnages.

One after another the bottlenecks were relieved until the trickle of pellets grew to a production rate of 4,000,000 tons per year, a quarter of a million more than the plant had been expected to produce. Still the pellet plant operators called for more ore. Finally, it became evident that the real bottleneck limiting the capacity of the whole E. W. Davis Works was the fine crushers.

There were eight big Symons cone crushers in four lines in the fine-crushing plant to reduce the rock received from the mine from three inches to three-quarters of an inch. In these machines the central wearing part, which oscillates to crush the ore, weighs 5,000 pounds. With constant use, it wears out and must be replaced about every two weeks after crushing about 15,000 tons of taconite. With eight machines in the plant, replacements were being made almost continuously and the production of one crushing line, or one-quarter of the plant capacity, was lost.

Fine crushing is an expensive operation, not only because of the cost of replacing the worn parts, but also because of the loss of production during the time the crusher is out of service for repair. In 1955–56 the maintenance department made a careful study of this operation with the idea of reducing the time that the crusher had to be "down" or out of service. The first time the worn crusher parts were replaced it took two days and nights, but by the time the Silver Bay plant had been operating a few months, this had been reduced to twelve hours. Before the first year was over, the "down time" had been cut to two hours, and later to only thirty-eight minutes. The maintenance men had done a marvelous job, but they could do no more.

When the fine-crushing plant was being designed, Mr. Linney

and his men believed that it was being planned for too small a capacity, and they attempted to persuade Fred Darner to enlarge the plant by including a fifth crushing line. But it could not be shown that the equipment was underdesigned for the production of 3,750,000 tons of pellets per year, and the design engineers were right. The fine-crushing equipment had ample capacity for the planned tonnage, but it turned out that the concentration and pelletizing units had excess capacity. This meant that unless the fine-crushing plant could produce more crushed rock, the rest of the plant could not operate at maximum efficiency. Mike Tourje, head of the industrial engineering department, and his assistants made an extended study of the situation and came up with the conclusion that a fifth crushing line would pay for itself very quickly in increased pellet production. Consequently, in 1956 the construction crew was back on the job, pouring cement and setting up steel.

As soon as the new fifth line of crushers went into operation early in 1957, there was plenty of crushed ore for the rod mills. In fact, more was now available than they could grind. To get the plant back in balance, the rod and ball mill capacities would have to be increased by about 15 per cent. After much discussion it was decided that the best way to do this was to lengthen the mills by two feet. Consequently, one section of mills was closed down and lengthened. In 1960 the task of lengthening all the mills was begun; it was a big job that took almost six months, shutting down one section at a time. When it was completed in 1961, certain bottlenecks then appeared in the concentration equipment, but with only small changes, the plant balanced up nicely, and all sections were working near maximum efficiency. These major changes and a host of minor ones gradually increased pellet production until a plant that was originally designed for 3,750,000 tons per year was by 1961 turning out almost 6,000,000 tons.[17]

How good were the pellets produced by the new plant? To find out, a method of measuring pellet quality was adopted which was similar to that used for testing the strength of coke. Fifty pounds of pellets, free of dust and breakage, were put into a three-foot drum, which was then rotated at a constant speed for a definite number of revolutions. The pellets rolled and tumbled upon one another and upon the lining of the drum, and any soft ones wore down and broke up. They were then removed and the amount of dust and breakage was measured.

This was quite a severe test. At the Mines Experiment Station, we had considered pellets acceptable if the amount of fines (dust and breakage) produced did not exceed 20 per cent. When the Babbitt plant started operating in 1952, the amount of fines produced by this tumble test averaged about 15 per cent. At Silver Bay such tests were made each day, and at first they averaged just under 20 per cent, which was not considered very good.

Each week there was a staff meeting to study pellet production, the idea being to improve the quality and make the type of pellets desired by the blast furnace committee of Armco and Republic.[18] Its members met frequently at Silver Bay, and they urged Reserve to produce higher-quality pellets, because these worked so much better in the blast furnaces. At one such meeting Mr. Kelley asked me, "How good can the pellets be made?" I ventured the answer, "Ten per cent fines after tumble test." A great groan went up from the pelletizing operators, but they continued to work on the processing and gradually improved the quality. By 1962 if the pellets contained more than 5 per cent fines after the tumble test, the operators were apologetic. This was much better than any pellets we had ever made at the Mines Experiment Station and better than I thought would ever be possible in a commercial plant.

The blast furnace committee also asked that the pellets before tumbling be made smoother and with as few small or broken pieces and as little dust as possible. The men in the pelletizing plant also tackled this problem, and by 1962 the pellets contained only about one-half of one per cent of fines smaller than 28 mesh.

At the Mines Experiment Station, pellets of various sizes had been made to demonstrate that the size was under the complete control of the operators. Before 1956, however, the preferred size was a pellet three-quarters of an inch in diameter. Studies by Republic and Armco furnace operators indicated that smaller pellets might work better. As a result, the size was first reduced to five-eighths of an inch in diameter and then was further cut to a half inch, then to three-eighths, and finally to eleven thirty-seconds of an inch. This small pellet has been made since 1960.

After it had been running almost a year, the E. W. Davis Works was officially dedicated on September 13, 1956. The more than six hundred people who attended the dedication saw a movie at Duluth showing the development of the whole project, and went by special train to Babbitt for a quick look at the mine and crushing plant. The train then took them over Reserve's railroad to Silver Bay, where the dedication ceremonies were held in one end of the big new concentrator building. The speaker for the occasion was Mr. Arthur S. Flemming, national director of defense mobilization, who also read a congratulatory letter from President Dwight D. Eisenhower. Mr. White and Mr. Sebald talked briefly, and everything went off smoothly under the direction of Edward Schmid, Jr., Reserve's public relations director. At first, Doc Kelley was somewhat scornful of the whole idea, saying it was like holding a wedding after the baby had arrived. But he got into the spirit of the occasion and acted the perfect host.[19]

Doc and Mrs. Kelley lived on the hill overlooking the lake and harbor at Silver Bay. They were interested in community affairs, and their home quickly became the gathering place for the older citizens of the village. Mrs. Kelley loved the informality of the area, and Doc was the most friendly and courteous person imaginable when at home. But in the plant he was impatient with any slackness or lack of attention to the requirements of the job.

On one occasion an engineer was to install a large strainer in the plant's main water line. This meant shutting off the water and closing down the entire plant for the time required to replace the strainer. To Doc nothing was more serious than interrupting the smooth operation of the plant, and careful plans were made to expedite the installation so it would take as little time as possible. The engineer and his crew went to work installing the big strainer, which weighed perhaps three tons. Everything went exactly according to schedule. The plant was ready to go back into operation again at the time planned, when it developed that the engineer had put the big strainer in backward. Although Mr. Kelley more or less controlled himself, the engineer will probably never forget the occasion, and he was quickly sent back to Cleveland.

The operation of the big plant gradually dropped into a routine, and it was plain that Mr. Kelley was beginning to feel that his work at Silver Bay was finished. Only when something out of the ordinary occurred did his interest revive. He was basically a builder and organizer, and the everyday routine did not interest him much. So it was no great surprise when he announced in 1958 that he was retiring.

Doc Kelley's contribution to the development of taconite processing in Minnesota can scarcely be overestimated. He took a little vest-pocket process out of a university laboratory and in one mighty step built a $185,000,000 machine,[20] staffed it, and put it into smooth operation, producing a higher-quality product at a higher production rate than had been anticipated. This done, he turned his masterpiece over to Mr. Linney, his successor as president, and quietly and gracefully retired.

Mr. Kelley left behind him one important piece of unfinished business. As we have seen, the contract with the Mesabi Iron Company which Reserve inherited from Oglebay Norton specified that Mesabi was to receive one-third of the net profit made by mining and processing Babbitt taconite.[21] In the 1950s the Mesabi company was inactive but still in existence. All that remained of the company's property was the contract with Reserve.

When Allen and Jackling prepared this agreement in 1939, they recognized that the determination of the profit made by a company operating the Babbitt property might involve complicated accounting and possible disagreements. For this reason they inserted an arbitration provision setting up a two-man board of arbitrators to decide all questions arising under the contract. If the two arbitrators could not agree, they were to name a third, and the majority could then decide. The decision of the referees was to be binding. Oglebay Norton recognized the difficulties that could develop in the interpretation of such a contract, and they asked Jackling, as president of Mesabi, to substitute a fixed royalty in place of the profit-sharing provision. No agreement was reached, however, and each year, beginning in 1954, Mesabi filed objections to Reserve's annual reports, claiming that Reserve had not complied with the provisions of the contract.[22]

In an attempt to settle the disagreement, Reserve in 1955 appointed Harrie Taylor, president of Oglebay Norton, as its arbitrator, and Mesabi appointed Louis Buchman. These two men settled a number of questions, but as the problems submitted to the arbitrators became more and more involved, Richard C. Klugescheid became Mesabi's representative and Bill Montague represented Reserve. These men found it impossible to agree on some of the knottier problems, with the result that on October 30, 1957, Wesley A. Sturges, former dean of the Yale Law School, was appointed as a disinterested third arbitrator. In 1957 and 1958 the three-man board held many meetings with Walter J. Milde and Eben H. Cockley of Jones, Day, Cockley, and Reavis, appearing as attorneys for Reserve, and Charles Pickett and L. B. Morey of Chadbourne, Parke, Whiteside, and Wolff, appearing as attorneys for Mesabi.

In 1957 in the midst of the proceedings, certain Mesabi stockholders became dissatisfied with the manner in which their directors were handling the discussions and themselves brought action against both Reserve and the then members of Mesabi's board. After a proxy fight in 1958 the new Mesabi directors refused to arbitrate, and Reserve then sued to compel Mesabi to resume arbitration. From 1957 through 1959 there was much legal tangling involving injunctions and actions in the United States district courts at Duluth and in Delaware. In 1959, Francis D. Butler of St. Paul was appointed by Mesabi as arbitrator to replace Klugescheid.[23]

The basic points in dispute were how to determine Reserve's cost of operation and a selling price for the pellets. The cost problem involved the question of amortization on Reserve's whole project. Mesabi insisted that under the terms of the contract Reserve could not charge against the cost of producing pellets any interest on the money borrowed for a new plant or for plant enlargement. The contract had been based on the theory that with minor changes the Babbitt plant would be a major unit in Reserve's commercial operation. Mesabi contended that, since Reserve had chosen to build new facilities as well as a power plant and a railroad, payments for interest and for the retirement of this new investment should not be included in computing the cost of producing pellets.

When it came to determining the selling price of the pellets, the issue became even more complex. Reserve's pellets belonged to Republic and Armco, each of which paid half the cost of operating the property and took half the pellets. Therefore, Reserve itself made no profit. The Lake Erie price might be used in computing a selling price for the pellets, but this calculation did not take into account the value of their improved structure.

Both Reserve and Mesabi were preparing for a protracted legal battle over these issues when the announcement was made that both parties had agreed on a settlement of the whole affair. Various suits were to be dismissed and various costs were settled, and Mesabi was to accept a fixed royalty of one dollar per ton of pellets in place of the profit-sharing provision in the original contract. In effect, Reserve pays Mesabi a royalty of one dollar per ton of pellets on the taconite mined at Babbitt under the Peters' lease. The agreement was ratified by the directors of Republic and Armco and by the stockholders of Mesabi, and on April 27, 1960, everything was signed and sealed and this incident in the history of taconite was closed.[24]

To the casual observer it may seem that Mesabi exacted a high toll for the small contribution it made to the success of the Reserve Mining Company or taconite generally. About the best that can be said for Mesabi is that it kept in force and turned over to Oglebay Norton and then to Reserve the Peters' lease on the taconite lands discovered by Christian Wieland and Peter Mitchell and later owned by the Williams-St.Clair-Mitchell heirs. Reserve had no use for the old machinery at Babbitt acquired under the Mesabi contract and had little use for the old mill buildings or for the houses in the old town of Babbitt that Mesabi had abandoned. Reserve did profit, however, by avoiding the mistakes that Mesabi had made when that firm planned, built, and for a short time operated the original Babbitt plant.

After the settlement in 1960, the Reserve taconite project moved ahead smoothly under the direction of Mr. Linney as president and Mr. Bryant as vice-president. Pellet quality and production rates continued to improve. Soon after Mr. Linney became president he appointed Ed Furness as his assistant. With Ed on the job, Bob felt that he could take time to look at some of the broader problems facing the taconite industry, including public relations which in earlier days had been so badly neglected by the natural ore producers.

As a result, he began to speak before various clubs and societies in the Duluth area, giving definite statistics and operating data on the Reserve project. In so doing, he dispelled some of the mystery that had grown up around it. (For example, a mine operator once asked me how much gold Reserve got out of its taconite. He could not believe that such an elaborate project had been built merely to produce iron.) Mr. Linney felt that Reserve had received from the state of Minnesota benefits of great importance—the taconite tax law and the permits that had been granted to use Lake Superior as a source of water and as a place to deposit taconite tailings. He undertook to acquaint the public with Reserve's contributions to the welfare of Minnesota, which had come about as a direct result of these privileges.

Ed Schmid and others from Reserve, Alex D. Chisholm and Joseph S. Abdnor of Erie, and Christian F. Beukema, president of Oliver, also spoke before various groups.[25] Even the author opened his mouth again. An attempt was made to answer questions about taconite honestly and completely so that everyone interested would understand the difference between taconite processing and the production of natural high-grade ores.

After I talked at one meeting I was asked, "How much does it cost Reserve to make a ton of pellets?" I answered, "Some place between ten and twenty dollars per ton." The questioner then asked, "How much do they sell them for?" I answered, "They don't sell them. The owners themselves use them." Realizing that this answer was not very helpful, I went on to ask what the questioner really wanted to know. He replied, "Are they making any money?" I told him that was something none of us would know until the steel companies which owned Reserve and Erie made their next moves. If the taconite plants were enlarged, the operations were profitable and pellets were competing successfully with other sources of ore. If production were curtailed, it would probably mean that the pellets were not able to compete.

As early as 1951, *Skillings*, the influential Minnesota mining magazine, had said astutely: "The extent of the development of a Minnesota taconite industry will depend upon the ability of Reserve Mining Co. and other taconite producers to reduce every item of operating cost to a minimum." In speaking to many groups, we tried to make clear why this was so by explaining the importance of the taconite tax law and giving actual investment, payroll, and tax costs, as well as tonnage figures. In my opinion, the mining industry has too long neglected to explain its problems and plans to the people, and much of the dissatisfaction with the industry on the part of Minnesotans stems from a lack of understanding. The taconite producers did not wish to make this mistake.

11 The Revolution of the 1960s

IN THE 1940S AND 1950S RESEARCH by the Mines Experiment Station and the various companies had demonstrated that taconite concentration could be accomplished in laboratories and pilot plants. Then Reserve's plant at Silver Bay and Erie's at Hoyt Lakes had proved the feasibility of large-scale production of pellets from Minnesota taconite. Until 1960, however, the ability of taconite pellets to compete commercially with rich, natural ores had not been conclusively established. The steel companies which owned Reserve and Erie had been using taconite pellets in their blast furnaces since these two big plants went into production in 1955 and 1957 respectively, but only as part of mixed furnace charges which also contained natural ores. No commercial furnace had as yet made iron entirely from taconite pellets. Only a full-scale test, in which a large furnace was charged with pellets alone, could prove their smelting quality. The crucial year which saw taconite passing this final test was 1960.

In order to appreciate fully the significance of what happened that year, it is necessary to recall the situation of the steel companies in regard to ore supplies. Little remained of the old, reliable, high-grade ores of the Mesabi Range, which had been the principal source of iron in the United States since the beginning of the twentieth century. Consumption of steel had been increasing steadily in the late 1950s and was expected to continue to do so as the population grew. Rich ore fields had been discovered in Labrador and Venezuela, and the deep St. Lawrence Seaway had made these ores, as well as others, readily available to United States steel plants.

Much of the natural Labrador ore was richer than the pellets made from taconite. In addition, as we have seen, the pellets were a high-cost product, and there was a question whether iron and steel could be made from them as cheaply as it could from high-grade natural ores. As a precaution, Republic, for example, had acquired control of large quantities of ore in Labrador and elsewhere, and early in 1960, when the controversy with the Mesabi Iron Company was at its height, serious doubts still existed about the future position of taconite in the industry. If the pellets could not compete with foreign ores, Reserve and Erie had made costly mistakes, and the future of Minnesota – indeed, of the whole Lake Superior ore-producing district – looked very dark.

Even as late as 1959, an informed observer, knowing the details and costs of producing taconite pellets, could not tell for sure whether they were competitive with foreign ores. The big unanswered question was: did the improvement in smelting properties that had been designed into the pellets overcome any advantage the foreign ores might have in chemical purity and cheapness of production? The question would be answered not in the Minnesota taconite plants, but by blast furnace operators in the nation's steel mills.

In the summer of 1960 Armco undertook the full-scale smelting test for which we had all been waiting. The Bellefonte Furnace at Middletown, Ohio, became available for test use, a good supply of pellets was on hand, and the furnace operators were free of commitments which made the use of other ores necessary. To thoroughly test the smelting qualities of taconite pellets, this large furnace was operated solely on them for several thirty-day periods.

At first no results of the test were announced publicly. Then in September, 1960, Thomas M. Rohan, Cleveland regional editor of *Iron Age*, secured permission to publish at least the bare facts in a wide-ranging article, entitled "Steelmakers Plan Pellet Plants to Head Off Foreign Inroads." He told his readers that a record "which stunned the industry" had "been achieved on long runs" of taconite pellets in Armco's blast furnace at Middletown. "A furnace rated about 1500 tons per day there," he went on, "has achieved daily production records of 2700 to 2800 tons of pig iron yield." Using taconite pellets alone, the Bellefonte Furnace had produced almost twice as much pig iron as normal!

Later the results were fully described in a technical paper by Kenneth R. Haley, superintendent of the furnace. He gave facts and figures on the use of the pellets compared with natural ores and concluded that under comparable conditions pellets increased production by 60 per cent while using 27 per cent less fuel. He said that by blowing more air into the furnace, production could be increased by 92 per cent! The key to the superior performance, Haley said, was the careful tailoring of the pellets, which "are most effective when used as 100 pct[.] of the ore burden." In other words, the only change made at the blast furnace to achieve these results had been to increase the capacity of the blowing equipment so that more air could be blown into the furnace. The pellets, as Doc Kelley had predicted, spoke for themselves. The test results conclusively demonstrated the reality of the theoretical advantages which we at the Mines Experiment Station thought the tailor-made pellets would have and the practical advantages of them recognized by Mr. White.[1]

In April, 1960, Mr. Linney had announced that Republic and Armco were considering expanding Reserve's Silver Bay plant, and on August 4, after the smelting test, he released the news that Reserve had begun construction of a $120,000,000 addition to the E. W. Davis Works which would increase its annual capacity to 9,000,000 tons.[2] The announcement that Reserve was enlarging its plant capacity by 50 per cent decisively answered the big question: the pellets were competing successfully with natural ores.

(I first heard the news that Reserve was to expand production while I was having my hair cut in the barbershop at Silver Bay. Later that day when I returned to the office, I was told in strictest confidence that orders had just been received from the

Cleveland office to enlarge the plant. If you want to know what is happening in Silver Bay, just go to the barbershop.)

Strangely, the press appears never to have fully appreciated the importance of Mr. Linney's statement that Reserve would expand its operations in Minnesota. The news was not discussed prominently, even in the Duluth or Two Harbors newspapers, although it radically changed the economic outlook of these cities. The absence of enthusiasm stood in notable contrast to the surge of excitement which followed the discovery of the Mesabi in the late nineteenth century; I am convinced, however, that Mr. Linney's announcement was news of equal importance.

Although Minnesota pioneered the development of magnetic taconite, the process was soon put into operation in other areas. Reserve and Erie were justifiably proud of their accomplishments and opened their doors at Babbitt, Silver Bay, and Hoyt Lakes to hundreds of visitors. The plants were always neat and ready for inspection. So much so, in fact, that one foreman was heard to complain, "My job here is to keep this place clean and after that to produce a few pellets, if I have time." [3] Because of the free exchange of technical information, there were no secrets in the taconite industry, and a number of other firms began to give serious consideration to the development of low-grade ore deposits. The Cleveland-Cliffs Iron Company had moved into pelletizing at Eagle Mills, Michigan, in 1955. Its engineers developed a successful process to produce good pellets from large Michigan deposits of jaspilite, a hard, low-grade nonmagnetic ore formation containing thin bands and flakes of specular hematite. The amount of this rock available in Michigan, Canada, and elsewhere, is comparable to the supply of Mesabi taconite; moreover, it is of a higher grade and is much easier to grind.[4]

In the early 1960s, Cleveland-Cliffs pioneered a number of new ideas in the production of high-grade iron ore pellets. The firm was not only the first to utilize the flotation method for the commercial concentration of hematite ores but also the first to use the double-firing technique proposed by Dr. Cooke for pelletizing hematite. At its Empire plant near Negaunee, Michigan, the company introduced autogenous grinding by the use of which low-grade ore can be pulverized commercially much finer than previously was considered possible — to 98 per cent -400 mesh. The pulverized material is then concentrated magnetically and pelletized. At the Mather Mine near Ishpeming, Michigan, the firm plans to complete by 1965 the installation of the grate-kiln method to pelletize the hematite ore from this underground property. In the future these new processes can and probably will be utilized in Minnesota and elsewhere.

Other companies have put taconite properties into production using processes similar to those developed in Minnesota. In 1961 and 1962 United States Steel opened taconite plants with combined capacities of over 9,000,000 tons in Quebec, Canada, and Atlantic City, Wyoming. By 1963, eight years after Reserve started up the world's first big plant, nine commercial pelletizing plants were in operation in the United States and five in Canada. Three more began shipping pellets early in 1964. As this is being written in June, 1964, pellet plants are also in operation in Sweden, Japan, and Peru, and others are under construction or in the planning stages in Norway, Italy, Japan, India, Australia, and Brazil. Since the Erie and Reserve projects were put into

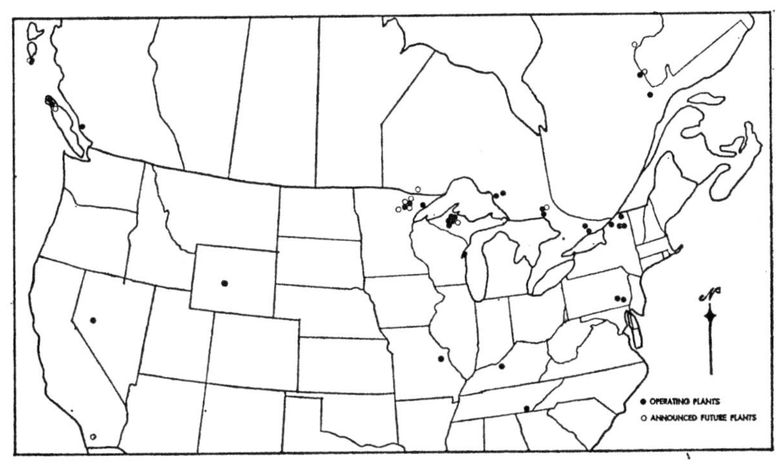

FIGURE 14. *Operating and announced taconite and taconite-type plants in North America (June, 1964). Based on information compiled by the American Iron Ore Association and the University of Minnesota Mines Experiment Station.*

operation, however, no new taconite plants have been built in Minnesota, although four possible ones are projected. Oglebay Norton and the Ford Motor Company have announced their intention to erect a plant to be named the "Thunderbird" near Virginia; United States Steel has placed a tentative target date of 1965 on a plant near Mountain Iron; the Hanna company has announced plans for a plant near Cooley on the western Mesabi; and Jones and Laughlin has obtained leases on taconite lands near Biwabik.[5]

The superior smelting qualities of the pellets have had even more far-reaching effects than we at the Mines Experiment Station had anticipated. The ability to control the size and shape of the ore particles charged to a blast furnace is bringing about a revolution in the iron and steel industry. Steelmakers no longer need to accept the raw material as the iron miners dig it out of the ground. Blast furnace operators can now get not only the analysis but also the size and shape of ore particles they want. The use of a uniform product of unvarying quality makes it possible for steel men to operate their furnaces at previously unobtainable levels of efficiency.

In a few short years the success of low-grade ore concentration and the demonstrated economy of carefully prepared blast furnace ore has brought about a complete re-evaluation of iron ore production and reserves. Instead of a shortage, a great surplus of high-grade natural ore now exists, and blast furnace operators who are willing to use direct ores are in a position to pick and choose among several sources. However, it is difficult to believe that the operators can long afford to charge raw ore into their furnaces. In the face of rising imports of foreign-made steels, United States producers find themselves in a highly competitive world market. They are naturally turning to the most economical ores, which, as we have seen, are the highly processed ones like taconite pellets, that can double furnace capacities. By 1963, over 26,900,000 tons of pellets were available to steel producers, representing 24 per cent of the total iron ore consumption in the United States and Canada as compared to about 12 per cent in 1960.[6]

The great improvement in smelting practices, brought about

by the availability of pellets made to specifications, has caused a corresponding revolution in the ore fields. Mines that have been producing high-grade ore for more than half a century are being abandoned, not because they are depleted, but for lack of a market, and other mines containing inferior ore of concentrating quality are being opened. In Labrador, where enormous deposits of rich ore lie waiting to be shoveled out of the ground, elaborate concentrating and pelletizing plants are being constructed to process nearby low-grade materials. Republic in 1961 announced that it had disposed of some of its high-grade natural Labrador ore, while retaining its interest in pelletized concentrate there and in Minnesota. On the Mesabi, Pickands Mather in 1961 shipped only 1,500,000 tons of natural ore from its remaining Minnesota reserves of over 40,000,000 tons, whereas it sent over 7,500,000 tons of taconite pellets to Eastern furnaces from Erie's Hoyt Lakes plant. There is every indication that this trend away from direct-shipping ores will continue, barring some totally unexpected scientific break-through in iron and steel production.[7]

The steel industry's acceptance of taconite pellets could be an event of great significance for Minnesota and the Mesabi Range. It promises permanent employment for residents of taconite towns. It implies that future production from the Mesabi can at least equal past production. Billions of tons of magnetic taconite may now be classified as an actual, rather than a potential, source of valuable ore, and still unestimated amounts of nonmagnetic semitaconite may be considered a probable, rather than a possible, supply. To members of the Mines Experiment Station, it means complete justification for the years of research which were supported by state funds. Acceptance of the pellets also has meaning for other parts of the world. Enormous beds of low-grade ore — now quite universally called taconite — exist on all seven continents, and pelletizing plants are now being projected all over the world.[8]

In view of the almost totally unexpected changes that have already occurred in the iron and steel industry, it is risky to predict what will happen in the future. However, certain lines of development seem clear. Such a revolution is certain to cause economic readjustments, hardships in some areas and prosperity in others, reaching from planning at the management level down through the steel plants to the workers in the mines. While expanded taconite operations create many new jobs, these are not necessarily in the same localities where mining was previously active. The production of a ton of pellets from taconite requires several times the number of men needed to produce a ton of natural open pit ore, although different skills may be required of the workers. In Minnesota during the 1963 season, for example, taconite operations employed more than twice as many man-hours of labor as all the natural ore mines in St. Louis County put together, although the natural ore producers shipped over 1,800,000 more tons of ore.[9]

The shift to the use of highly processed ores directly benefits ore-producing districts because it brings them payrolls of about $3.25 per ton in place of the $1.50 per ton they receive from natural open pit ores. According to the Minnesota Natural Resources Council, in 1961 the total taxes plus payrolls left in Minnesota per ton of taconite pellets shipped was over 50 cents greater than that from each ton of natural ore shipped. The trend also benefits the steel industry by reducing costs, thus making the producers better able to meet foreign competition and the growing inroads of aluminum and plastics.[10]

Nationally, one aspect of the shift to the use of both highly processed and foreign ores should be considered — the availability of the very large amounts of iron ore which would be needed in a wartime emergency. In 1942, early in World War II, iron ore shipments to the steel plants of the United States reached a total of almost 106,000,000 tons compared to prewar 1939 shipments of about 55,000,000 tons. Of this enormous tonnage, 66 per cent came from the Mesabi Range. The great Hull-Rust-Mahoning group of mines at Hibbing increased production to the unbelievable total of 27,000,000 tons in one shipping season. Perhaps the next war will be over in a few days, and there will be no need for guns, tanks, battleships, and cargo vessels, but it should be recognized that in spite of taconite the Mesabi Range cannot again be counted on to answer an emergency call for a large, rapid increase in iron ore production. Only big open pit mines can greatly step up ore shipments virtually overnight. Moreover, although many millions of tons of ore of smelting quality remain on the Mesabi, the best of this ore is being mined at the rate of 40,000,000 or 50,000,000 tons a year.[11]

As we become more and more dependent upon taconite for a domestic ore supply, it must be remembered that it will take from three to five years, even with a crash program, to bring a new 10,000,000-ton taconite project into production. It will also take thousands of tons of steel for construction and machinery at a time when such materials may be badly needed for military purposes. Taconite is a mighty giant, but it cannot completely take the place of the great Mesabi open pits.

Much additional geological exploration is necessary to determine even the full potential of the remaining ore resources of northern Minnesota. All the mining activity and tonnage statistics of the Mesabi have been confined to a relatively narrow band of iron formation where it comes to the surface to form this great range. However, there is reason to believe that truly impressive quantities of iron ore exist in one form or another at great depths. A proposal has been made to drill a few holes, a mile or more in depth, through the overlying rock in the area between the Mesabi and Lake Superior. Compared to other public works, this would not be an expensive undertaking, and it would provide information that would make really long-range planning possible. (See Figure 1.)

We know that the taconite dips below the surface and continues for at least a short distance down the slope from the Mesabi outcrop toward Lake Superior, but mining it for any distance would mean the removal of hundreds of feet of surface rock. This would have been economically unfeasible a few years ago, but with today's improved drilling methods and atomic blasting, the mining of deep ores seems more possible.

Experts associated with the Atomic Energy Commission have already given some thought to uncovering deep bodies of ore. Assuming that the taconite band continues for the length of the Mesabi Range, each mile down the slope toward Lake Superior is estimated to have a potential of five to ten billion tons of high-grade concentrate. (The total amount of ore mined on the Mesabi since its discovery is less than three billion tons.) Experts on the underground mining of taconite have been studying ways of digging beneath the great mass of rock which covers the Mesabi iron formation south of the range. This project is feasible today if the rock capping is sufficiently solid and strong to make possible safe mining beneath it. The characteristics of the rock can be determined only by deep drilling.[12]

It would seem that the amount of iron potentially available in the Mesabi formation is sufficient to justify the term "practically inexhaustible," but there is still more. Magnetometer records published by the United States and the Minnesota geological surveys point to the possibility that Minnesota taconite extends westward nearly to the state border. (See Figure 15.) Little drilling has been done in that area. The western part of the state is deeply covered by glacial drift, and at present no estimate can be made of the extent of the taconite beds.

It is expected that ore consumption will expand more or less

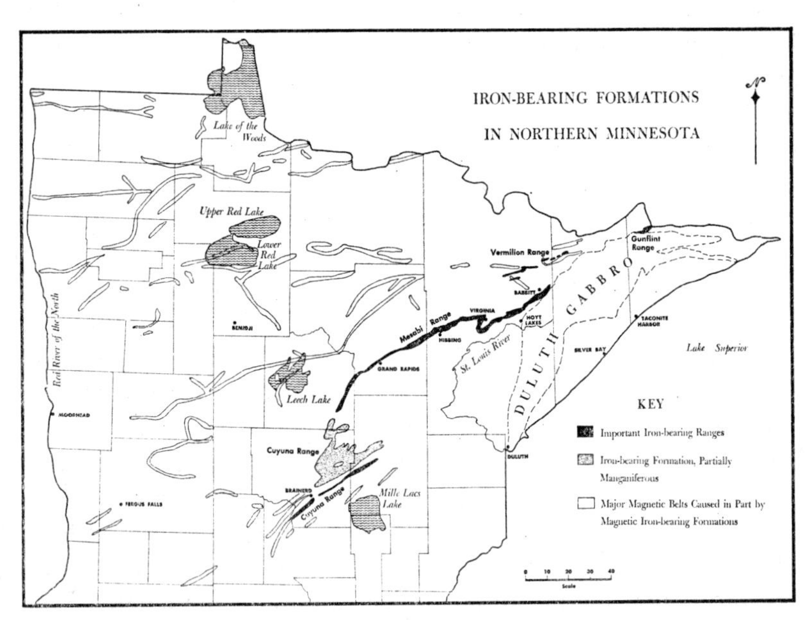

FIGURE 15. *Iron-bearing formations in northern Minnesota. Based on information prepared by the Minnesota Geological Survey.*

proportionately with predicted increases in population. In recent years, the center of population has been shifting westward, which suggests that in the future Chicago and St. Louis will produce more steel and require more iron ore. The natural source of pellets for the Chicago area is the Lake Superior district, and it seems likely that Chicago will become a large market for Lake Superior pellets. With a little assistance, Minnesota should be able to capture a substantial share of that market. If a new canal were dug across Michigan's Upper Peninsula, Minnesota's transportation advantages over Labrador and other foreign ore and pellet sources could be great.

It would seem that the pioneering period of taconite processing is drawing to a close. Obviously, however, the story of this new industry is far from finished. New pages are being added almost weekly as research continues at the Mines Experiment Station and elsewhere to develop processing methods for nonmagnetic semitaconites and other offgrade ores. The station and various firms are investigating magnetic roasting, flotation, and other processing methods, and there is every reason to believe that commercial plants will soon be concentrating and pelletizing this material in localities where conditions are favorable.[13]

Work in improving the quality of taconite pellets still progresses. In the Silver Bay office of the president of Reserve a drawer, when opened, displays about twenty-five different kinds of pellets which can be made from taconite concentrate. Some are the size of tennis balls, some the size of blueberries. Some are cylindrical, some are square, some are glazed in different colors. Some are practically pure iron, others contain enough lime to make them self-fluxing. Some are so hard they can be broken only under extreme pressure. Some have high porosity; others are dense. They may be composed of hematite, magnetite, or metallic iron. Any of these pellets can be put into large-scale production if the cost is warranted by improved blast furnace performance. In 1964 blast furnace operators wanted pellets about eleven thirty-seconds of an inch in diameter, with a tumble test rating of less than 6 per cent, analyzing 64 per cent iron and just under 8 per cent silica. Next year they may want something different, and it can be made for them. Like automobiles, new models can be produced each year.[14]

As this is being written, it is evident that pellets made from highly processed iron ore concentrates have passed every test, but in Minnesota taconite faces a new hurdle. In spite of the taconite tax law enacted in 1941, some present-day steel company officers are not satisfied with the iron ore tax situation in Minnesota and have been reluctant to invest there the capital required for taconite plants. Whether they are justified in this attitude, the reader must judge for himself. Nevertheless it is a fact that many companies have built and are building plants beyond Minnesota's borders, not because of any shortage of taconite in the state, but because other states and nations with as good or better taconite offer a "tax climate" more to the companies' liking.

The trouble seems to lie in the special tax laws which apply only to the mining of iron ore and taconite in Minnesota.[15] Since the production of pellets is a manufacturing process, potential taconite producers desire to have the taxes on their industry geared to those of other manufacturing industries. They do not like the idea that taconite production may be singled out for tax increases as high-grade ores were in the past. The producers do not suggest that taconite taxes be reduced or fixed at a given level. They ask only that their taxes be increased proportionately as those on other industries are increased in the future.

Recognizing the justice of this request, a group of legislators from the mining area proposed to assure what came to be known as "fair tax treatment" for taconite by introducing in the 1961 legislature a bill for a constitutional amendment containing these provisions. The measure aroused considerable controversy, especially in the Democratic Farmer Labor party, and was never brought to a vote. In the 1963 legislative session a similar bill was passed after a time limit of twenty-five years had been placed on its provisions.[16] It will be on the ballot in the 1964 general election. To pass, the amendment must receive the affirmative vote of a majority of all those voting. It has been endorsed by both major political parties, many labor, business, professional, and other groups, and a bipartisan citizens' committee was appointed by the governor to enlist public support for the measure.[17] About the only organized opposition has come from a small but vocal group of DFL legislators.

The future of the taconite industry in the state where it was born may depend upon the outcome of the vote on this amendment. If it passes, taconite may take root and flourish, and a new chapter in the history of the Mesabi will be written. It is estimated that there is enough taconite in Minnesota to supply ten large plants — plants that could provide employment for all the people of the iron ranges.

There is a sufficient supply of taconite — of the types that we have learned to mine and use — in the narrow outcropping bands of the Mesabi and Gogebic ranges alone to supply amply the needs of the present. Future generations will develop methods of mining and processing these and other ores as undreamed of today as the present ones were when Peter Mitchell roamed the Giant's Range and found his mountain. The iron minerals are there in various forms, and it only remains for future scientists to learn how to use them economically. We know now, as Peter Mitchell did not, that highly processed low-grade ores can compete favorably with natural high-grade ores. We can be sure that future explorers and scientists will take it from there.

Appendix 1

Taconite Technology and Pricing

IRON (FERRUM, CHEMICAL SYMBOL FE) is a gray strongly magnetic metal with an atomic weight of 56. It combines readily with oxygen (chemical symbol O, atomic weight 16) to form ferric oxide or hematite (Fe_2O_3 – 70 per cent Fe) and ferrous oxide or wustite (FeO – 77.7 per cent Fe), both of which are nonmagnetic. Ferric oxide as it occurs on the Mesabi Range is ordinarily combined chemically with water to form the mineral geothite or limonite ($HFeO_2$). This is the principal iron mineral of the soft, red-brown, earthy ore that has been mined on the Mesabi for many years. Ferric oxide (Fe_2O_3) also combines with ferrous oxide (FeO) to form the mineral magnetite (Fe_3O_4 — 72.4 per cent Fe). This hard, black or gray, glassy, strongly magnetic oxide is the principal iron mineral in the magnetic taconite of the eastern Mesabi.

Ferrous oxide does not occur as an independent mineral but combines with many other oxides to form such nonmagnetic minerals as cummingtonite, fayalite, hedenbergite, and other complex chemical compounds of iron, silica, calcium, magnesium, aluminum, and so forth, which can be grouped under the term silicates. These complex minerals seldom contain as much as 30 per cent iron and are not useful as a source of high-grade iron ore. For example, cummingtonite – the most abundant of these silicates occurring on the east Mesabi – in its pure state contains only about 20 per cent iron.

The taconite of the east Mesabi is a hard, dense rock, composed largely of an intimate mixture of silicates with small magnetite crystals. Only the magnetite crystals contain sufficient iron to be of importance, and these constitute only about one-third of the weight of the taconite. Magnetite is only 72.4 per cent iron, and therefore east Mesabi taconite contains only 20 to 25 per cent iron in the magnetic form (72.4 divided by 3 equals 24.1). The silicates may add as much as 5 to 10 per cent additional iron to the taconite, but most of it must be discarded as tailings in any processing operation that produces high-grade iron concentrate. For this reason the total iron content of the taconite is not of primary importance. It is only the iron that occurs as the mineral magnetite with which magnetic concentration processes are concerned.

Since the chemical analysis of taconite does not indicate the percentage of iron which is present in the magnetic state, the Mines Experiment Station's staff developed a method of securing the magnetic iron assay by using the magnetic tube machine described in Chapter Two. The magnetic iron assay, or Mag. Fe as it is called, differs from the total iron assay, or total Fe, which can be determined by chemical analysis. The Mag. Fe in taconite is always lower than the total Fe. The difference depends upon the amount of hematite and other nonmagnetic iron compounds that are present in the sample.

The magnetic iron assay is determined by grinding the sample to be analyzed very fine and then concentrating it in the magnetic tube. The concentrate will contain virtually all the magnetite present in the sample, and the tailings will contain virtually all the nonmagnetic minerals. From the weights and assays of these products the magnetic iron assay can be computed.

For example, if a ten-gram sample of finely pulverized -150 mesh taconite, analyzing 31 per cent total iron, is concentrated in the tube and the weight and total Fe assay of the concentrate and tailings are determined, the results and calculations can be arranged as follows:

PRODUCT	WEIGHT IN GRAMS		TOTAL IRON ASSAY		GRAMS OF IRON
Feed	10	×	31%	=	3.10
Concentrate	4	×	64%	=	2.56
Tailings	6	×	9%	=	.54

The 2.56 grams of iron in the concentrate represent all the iron existing as magnetite in the original sample. To determine the magnetic iron assay, one then divides 2.56 by 10 (the weight of the original sample) to get .256 or 25.6 per cent — the per cent of the original sample that is iron in the form of magnetite. This is usually

written as % Mag. Fe, and the above calculations are usually expressed as percentages:

PRODUCT	% WEIGHT	% IRON	PERCENTAGE UNITS
Tube Feed	100	31	3100
Concentrate	40	64	2560
Tailings	60	9	540

2,560 divided by 100 = 25.6% Mag. Fe.

The magnetic iron assay can also be determined if the assays of the feed, the concentrate, and the tailings are known but their weights are not known. To do this an algebraic formula is derived as follows:

$$(1)\quad \frac{\text{Feed assay} - \text{Tailings assay}}{\text{Concentrate assay} - \text{Tailings assay}} \times 100 = \%\ \text{Wt. Conct.}$$

$$\frac{31 - 9}{64 - 9} \times 100 = \frac{22}{55} \times 100 = 40\ \%\ \text{Wt. Conct.}$$

Then 40 times 64 equals 2,560, divided by 100 gives 25.6% Mag. Fe.

Frequently, the calculation as it is used commercially is still further simplified. It is obvious that the iron assay of the concentrate made in the tube depends upon grinding the sample fine enough to liberate all the particles of magnetite from the attached particles of the silicates. If this is done, the concentrate will be pure magnetite assaying 72.4 per cent iron and the tailings will contain all the non-magnetic iron. In order to determine the magnetic iron assay, it is only necessary to know the total iron assay of the original sample and the total iron assay of the tailings. The assay of the concentrate is assumed to be 72.4 per cent, which would be the case if the sample were pulverized sufficiently fine. The magnetic iron assay is then computed as follows:

$$(2)\quad \frac{\text{Sample assay} - \text{Tailings assay}}{72.4 - \text{Tailings assay}} \times 72.4 \text{ gives magnetic iron assay}$$

This is usually written:

$$(3)\quad \frac{\text{Feed} - \text{Tails}}{72.4 - \text{Tails}} \times 72.4 \text{ equals } \%\ \text{Mag. Fe}$$

Or to use the example above:

$$(4)\quad \frac{31 - 9}{72.4 - 9} \times 72.4 \text{ equals } 25.1\%\ \text{Mag. Fe}$$

The resulting figure compares with the 25.6 per cent figure obtained by the more accurate method involving both weights and assays. The shorter method, however, is sufficiently accurate for commercial metallurgical control requirements.

Formula 1 above is called the "two-way-split formula." It is useful in many metallurgical calculations. In a large taconite plant, the ore may be entering a machine or group of machines at the rate of several hundred tons per hour. In these machines it may be divided into two products: concentrate and tailings. Since it is often important to know the proportion of the feed that is entering each product, it is only necessary to secure small, accurate samples of the feed and the two products being made from it, and analyze them for total iron. Then by using Formula 1 the percentage by weight in each product can be computed.

An example of the use of this formula may be found in Chapter Four, note 10, where the metallurgical efficiency (% Mag. Fe recovery) was determined when tailings of different magnetic iron analysis were discarded by the equipment in the processing flow sheet. If the crude taconite from the mine averaged 24 per cent Mag. Fe and the desired grade of concentrate was to contain 65 per cent Mag. Fe and the tailings varied from 3 to 10 per cent Mag. Fe, the metallurgical efficiency (or the per cent of Mag. Fe recovered) could be calculated as follows:

$$(5)\quad \frac{24-3}{64-3} = \frac{21}{61} = .344 \text{ or } 34.4\% \text{ weight recovery of concentrate}$$

$$\frac{34.4 \times 64}{24 \times 100} = 92\% \text{ Mag. Fe recovery or } 8\% \text{ loss}$$

Similarly if the tailings assayed 4 per cent Mag. Fe, the percentage of Mag. Fe recovered would be 89 per cent, or an 11 per cent loss. If the tailings assayed 10 per cent Mag. Fe, the percentage of Mag. Fe recovered would be 70, or a 30 per cent loss.

For efficient magnetic taconite mining and processing, assays of the various products are usually reported as the percentage of magnetic iron in a dry sample. The ore may be damp or flowing with water but only its dry weight is used in computing the assay. However, on the final product to be shipped and sold, total iron assays are reported, based on the wet or "natural" weight of the ore. The natural iron assay may be secured by subtracting the per cent of moisture in the ore from 100 and then multiplying the result by the dry iron assay. Ore that contains 10 per cent moisture and assays 60 per cent iron (dry) will thus assay 54 per cent iron (natural).

The reason for the use of total iron assays can be readily appreciated when it is recalled that the total iron assay of the pellets made from taconite is always considerably higher than their magnetic iron assay because pelletizing converts most of the magnetite to hematite. The producer expects to be paid for every pound of iron he ships. Natural weights are used because the railroad and steamship operators must be paid for the tonnage they transport, whether it is ore or water.

In published statistics both dry and natural analyses are frequently shown. This is largely a matter of convenience. If natural iron assays are used with natural weights and dry iron assays are used with dry weights, the actual tons of iron are the same. For example, if a boatload of 10,000 tons of ore has a moisture content of 10 per cent and an analysis of 54 per cent iron natural, then 10,000 times 0.54 or 5,400 tons of iron were shipped. Similarly, if the dry weight of the shipment was 9,000 tons and the dry analysis was 60 per cent, 5,400 tons of iron were shipped.

It is also customary in the iron ore industry to report and record weights in gross or long tons. A standard or net ton contains 2,000 pounds, but a gross ton is 12 per cent larger and contains 2,240 pounds. One per cent of this, or 22.4 pounds, is called a "unit," and a gross ton of ore assaying 60 per cent iron is said to contain 60 iron units. Obviously the number of iron units in a shipment is the same whether it is computed from dry weights and dry iron assays or from natural weights and natural iron assays. The unit figure is used in price or value computations.

The Lake Erie Base Price, which is established each year for the various grades of iron ore, is determined by the early season sales as announced in the trade journals of the industry. There has been much argument over the years about this pricing system, because it not only establishes the market price for immense tonnages of ore each year, but it is used in computing the taxable profits realized by mining companies. Although a given shipment of ore may not actually be sold at the computed Lake Erie figure, it is the first, or asking price, of the producer in his negotiations. The base figure is established each year for particular grades and qualities of ore; if the ore is better than the base, the price is proportionately higher; if lower, the price is lower. In theory it is possible to determine the price of any quality of ore delivered at lower Great Lakes ports, but anyone experienced in selling iron ore knows that, while the computed Lake Erie price is useful for a number of purposes, the sale of ore — like the sale of anything else — is made at the best price the seller can get.

While the method of computing the Lake Erie base has changed many times over the years, in general the ore is first classified according to its structure and physical qualities. The hard ores are called "Old Range" and receive a premium, and the soft, earthy ores are called "Mesabi" regardless of where they are mined. Recently a new classification has been established for pellets; they are listed as a separate product.

Ore is also classified as either "Bessemer" or "Non-Bessemer" depending upon the amount of phosphorus it contains. If it contains more than .045 per cent phosphorus, it is classified as Non-Bessemer; if less than .045 per cent it is classified as Bessemer and carries a premium. The lower the phosphorus, the higher the premium. The reason for this lies in the smelting processes used by steelmakers. The widely used open hearth process removes the phosphorus contained in the ore, but the more rapid and less expensive Bessemer process does not. Therefore only ores which are low in phosphorus are used to make Bessemer steel.

The following table shows the fluctuations in the base price of Old Range Bessemer Ore for some years from 1913 through 1960, but it does not show the changes in the premiums that could be added for any particular shipment.

OLD RANGE BESSEMER ORE

YEAR	BASE PRICE*	YEAR	BASE PRICE	YEAR	BASE PRICE
1913	$4.40	1919	$6.45	1925	$4.55
1914	3.75	1920	7.45	1930	4.80
1915	3.75	1921	6.45	1935	4.80
1916	4.45	1922	5.95	1940	4.75
1917	5.95	1923	6.45	1945	4.95
1918	6.40	1924	5.65	1950	8.10
				1960	11.85

The announced 1964 Lake Erie Base Price for Lake Superior ore per gross ton (51.5 per cent Fe natural) delivered to lower Great

*From 1913 through 1924 the iron base was 55 per cent; after 1924 it was 51.5 per cent.

Lakes ports, as computed from *Skillings Mining Review* for April 18, 1964, is shown below.

1964 LAKE ERIE BASE PRICES

CLASSIFICATION	BASE PRICE PER TON	PER IRON UNIT
Mesabi Non-Bessemer	$10.55	20.48 cents
Mesabi Bessemer	10.70	20.78 cents
Old Range Non-Bessemer	10.80	20.97 cents
Old Range Bessemer	10.95	21.26 cents
Pellets	12.98	25.20 cents

In the above table it will be observed that in 1964 Bessemer ore received 15 cents more per ton than Non-Bessemer ore, and Old Range ore received 25 cents per ton more than Mesabi ore. For purposes of comparison, the base price of 51.5 per cent Fe pellets has been computed, although no pellets this low in iron analysis are shipped. The calculation indicates, however, that pellets receive a structure premium of about $2.00 per ton over other Lake Superior ores, because of their better smelting qualities.

Appeudix 2

Reduction Processes

PELLETS ARE THE MOST DESIRABLE FORM of ore thus far developed, because of their physical properties – their spherical shape and high microporosity. In the iron and steel industry the term reduction generally means a reduction in the amount of oxygen in the ore, or more explicitly, its partial or complete deoxidation. Reduction is effected by bringing into physical contact with the iron oxide, under properly controlled conditions, elements or chemical compounds that have a greater affinity for oxygen than has iron. The important reducing agents in use are hydrogen and carbon. Hydrogen, being a gas, is highly mobile and its molecules can search out the molecules of iron oxide. Since hydrogen has a stronger affinity for oxygen than oxygen has for iron, at elevated temperatures the iron gives up its oxygen to the hydrogen. Carbon, however, is a solid even at elevated temperatures and, because of its low mobility, cannot freely search out the oxygen in the iron oxide, which is also a solid. However, when carbon is partially oxidized, it forms carbon monoxide, a highly mobile gas that can penetrate the iron ore and search out the oxygen. Although many complex chemical reactions are involved, these two gases, hydrogen and carbon monoxide, are the agents in universal use for iron ore reduction. The latter is by far the most important, because of the widespread use of the blast furnace to reduce iron ore to pig iron.

Natural gas, largely methane (CH_4), is not a satisfactory reducing agent for iron oxides. However, if this gas is partially oxidized, it becomes a mixture of carbon monoxide and hydrogen and is a very effective reducing agent. This is accomplished by burning the gas with insufficient air or by mixing it with steam and heating to a high temperature. The reaction can be written:

$$CH_4 + O = CO + 4\ H \text{ or } CH_4 + H_2O = CO + 6\ H$$

Chemical reactions between solids and gases proceed rapidly at the surface of the solid but become progressively slower as deeper penetration of the solid is required. For this reason, the porosity of the solid is very important for rapid reduction, since high porosity increases the surface that is exposed to the reducing gas. The molecules of which the reducing gases are composed are submicroscopic; therefore, the more numerous the micropore spaces in the solid, the greater the surface exposed and the more rapid the reduction. High microporosity is thus a desirable characteristic of iron ore, and it is one of the reasons why taconite pellets reduce so rapidly. A properly made pellet will absorb water like a blotter, because of its porosity and the large surface exposure.

Taconite pellets are fired in a furnace below fusion temperatures in order to retain their microporosity, for fusion closes up the microscopic pores. As a rough illustration, a molecule of carbon monoxide in a blast furnace can enter a pellet as easily as a fly can get into a house with its windows open. Fusion closes the windows.

BLAST FURNACE SMELTING is the oldest and by far the most important method used for reducing iron ore. The impure metal produced is called pig iron. A modern blast furnace will produce a ton of molten pig iron per minute, some as much as two tons per minute. To make steel from pig iron, the impurities must be removed by further complex processing.

A blast furnace is an enormous stacklike structure made of steel lined with firebrick. It may be a hundred feet high and thirty or more feet in diameter inside the brickwork. Equipment is provided to keep the stack full to the top at all times with the furnace charge, which consists of many hundred tons of iron ore, coke, and limestone. At the bottom of the furnace shaft is the hearth, where the molten iron and slag collect. Near the bottom of the hearth are two openings that can be plugged or opened by the operators. The lowest opening is the taphole through which the molten pig iron is removed. A few feet above this is the "cinder notch" through which the molten slag is flushed. Above this are openings around the hearth, called "tuyères," through which preheated air is blown into the furnace.

After ignition at the tuyère level, the coke in the charge near these openings burns at a high temperature; the hot products of combustion pass upward through the charge, heating it and causing various important chemical reactions. Near the tuyères the temperature is so high that everything in that zone melts except the coke. Until it burns this fuel remains hard and strong enough to support the many tons of furnace charge in the stack above it. As the materials melt and the coke burns away, the charge in the furnace shaft settles, bringing new material to the tuyère zone.

The many chemical reactions that occur in the furnace shaft are determined by the temperature and by the nature and concentration of the gases produced by the combustion of the coke. These gases are principally carbon monoxide, carbon dioxide, and nitrogen. The nitrogen is relatively inert and serves largely to carry heat into the upper zones of the shaft. The carbon monoxide is very reactive, however, and combines with the oxygen in the ore to produce metallic iron and carbon dioxide. The complete reaction may be written: $4\ CO + Fe_3O_4 = 4\ CO_2 + 3\ Fe$. The CO_2 gas passes up through the charge and is discharged from the top of the furnace. The iron absorbs some carbon from the coke and melts to form pig iron. The earthy impurities in the ore and coke ash combine with the lime in the fluxing stone to form molten slag. The molten iron with the lighter slag floating on top collects in the furnace hearth just below the tuyères. Every few hours the slag notch is opened and the slag is flushed out. It flows into a ladle and is taken to the slag disposal area.

A few times a day the operators open the taphole, and the molten pig iron flows into an iron ladle. It is then cast into pigs weighing about one hundred pounds each, or it is carried in the molten state to a steelmaking furnace, where the impurities and most of the carbon are removed to make steel.

To produce a ton of pig iron, approximately the following weights and raw materials are required: 1.75 tons of iron ore, 1.0 tons of coke, .5 tons of limestone, and 4.0 tons of air. The weight of air required is greater than the combined weights of all of the other materials entering the furnace. Spherical pellets when charged to a blast furnace leave large and uniformly distributed void spaces, thus allowing the free circulation of the furnace gases and more rapid and complete reduction.

Magnetic Roasting. Much of the iron formation of the Mesabi Range west of the town of Mesaba is of the nonmagnetic type; that is, the principal iron mineral is hematite. Taconite plants like those operated in Minnesota by Erie, Reserve, and United States Steel will not produce concentrate from this material. It is possible, however, by the removal of a small amount of oxygen to convert nonmagnetic or semitaconite to magnetite. After this has been done, the material can be concentrated in existing taconite plants with some simple modifications.

The conversion of the nonmagnetic to the magnetic oxide of iron occurs rapidly at elevated temperatures and in the presence of such reducing gases as carbon monoxide and hydrogen. Conversion occurs at temperatures as low as 600 or 700 degrees Fahrenheit, but for rapid reduction temperatures of something over 1000 degrees Fahrenheit are used. After the ore is heated and the hot reducing gas is brought into contact with it, the chemical reactions may be written: $3\ Fe_2O_3 + CO = 2\ Fe_3O_4 + CO_2$. This shows that one hundred pounds of hematite react with 5.8 pounds of carbon monoxide to form 96.7 pounds of magnetite. The reaction with hydrogen may be written: $3\ Fe_2O_3 + 2\ H = 2\ Fe_3O_4 + H_2O$. This shows that one hundred pounds of hematite react with .40 pounds of hydrogen to form 96.7 pounds of magnetite. However, care must be taken with these reactions. If the temperature is too high or the concentration of reducing gas too strong, the reactions will produce FeO rather than Fe_3O_4 and FeO is a nonmagnetic oxide.

DIRECT REDUCTION has attracted the attention of investigators for a hundred years or more as a substitute for blast furnace smelting. The interest in this process lies in the fact that metallurgical coke is not required, and cheaper fuel that can be gasified may be used.

Modern taconite processing plants can produce high-grade concentrate, but at best their final product cannot exceed 72.4 per cent iron. The 27.6 per cent that they cannot remove is oxygen. The removal of this oxygen is called direct reduction or metalizing, because after all the oxygen has been removed, the iron is in the metallic state, although not necessarily melted.

In the magnetic roasting process previously described, only a small part of the oxygen in the ore is removed. If this operation is carried further using higher temperatures, higher concentrations of reducing gases, and longer contact time, metallic iron will be the final product. This reaction can be written: $Fe_3O_4 + 4\ CO = 3\ Fe + 4\ CO_2$, indicating that one hundred pounds of magnetite react with 48.3 pounds of carbon monoxide to form 72.4 pounds of metallic iron. Or if hydrogen is used: $Fe_3O_4 + 8\ H = 3\ Fe + 4\ H_2O$, indicating that one hundred pounds of magnetite react with 3.5 pounds of hydrogen to form 72.4 pounds of metallic iron.

Since these reactions proceed rapidly below the melting point of the ore, the iron produced is in the form of a spongy mass. It can be compressed either hot or cold into a more solid form, such as briquets, which can be used to some advantage in present-day steel plants.

The direct reduction process, however, removes only the oxygen, leaving any impurities that were present in the ore — such as silica or silicates — in the iron sponge. These can be removed by melting and fluxing in electric furnaces and in other ways, but preferably the impurities should be removed as completely as possible before the ore is metalized. Ores that are very pure or that can be concentrated to a high degree of purity are most desirable for direct reduction.

Such pure concentrate can be made from Minnesota taconite but is much more easily made from ores found in other districts such as New York, Missouri, Pennsylvania, Canada, and South America. The ore from the Adirondacks region in New York, for example, when processed like Minnesota taconite, produces concentrate assaying 72 per cent iron and containing less than one per cent of earthy impurities. Although experiments on the direct reduction of this and similar ores have been under way for many years, up to 1964 the process has not been shown to be competitive with blast furnace smelting in the United States. Direct reduction is being used to some extent in Norway, Sweden, and Mexico in localities where coking coals are expensive and electric power is comparatively cheap.

The demand of the steel producers is now, and as far as can be seen in the future will be, for high-grade ore that has been put into good physical condition for blast furnace smelting. This ordinarily means fine crushing and concentration followed by agglomeration. The blast furnace will not smelt taconite concentrate without agglomeration but blows it away as dust. If some new type of furnace is developed which is especially adapted to the smelting or metalizing of fine ore concentrate, it might be possible to avoid the agglomerating step in taconite processing with a subsequent saving in cost of approximately $2.00 per ton of ore. At the present time, it seems realistic to look forward to the day when fine taconite concentrate will be smelted or metalized in taconite plants and metallic iron, rather than iron oxide, will be shipped down the lakes for further processing and fabrication.

Appendix 3

Agglomerated Ore Shipments

LAKE SUPERIOR DISTRICT, 1918–1963

YEAR	COMPANY*	TONS	PRODUCT
1918	Mesabi Syndicate	1,840	Sinter
1922	Mesabi Iron	10,676	Sinter
1923	Mesabi Iron	83,525	Sinter
1924	Mesabi Iron	43,689	Sinter
1929	Mesabi Iron	11,041	Sinter
1949	Erie	15,756	Pellets
1950	Erie	62,087	Pellets
1951	Erie	137,607	Pellets
1952	Erie	93,527	Pellets
	Reserve	12,861	Pellets
1953	Erie	211,240	Pellets
	Reserve	245,643	Pellets
	U.S. Steel	104,464	Sinter, Nodules
1954	Erie	184,314	Pellets
	Reserve	344,183	Pellets
	U.S. Steel	360,363	Sinter, Nodules
1955	Erie	189,829	Pellets
	Reserve	333,352	Pellets
	U.S. Steel	632,195	Sinter, Nodules
1956	Erie	230,999	Pellets
	Reserve	3,909,113	Pellets

1956	U.S. Steel	676,797	Sinter, Nodules
	Cleveland-Cliffs	35,000	Pellets
1957	Erie	262,094	Pellets
	Reserve	5,421,205	Pellets
	U.S. Steel	664,243	Sinter, Nodules
	Cleveland-Cliffs	405,520	Pellets
1958	Erie	2,694,450	Pellets
	Reserve	4,994,174	Pellets
	U.S. Steel	732,876	Sinter, Nodules
	Cleveland-Cliffs	627,273	Pellets
1959	Erie	4,088,156	Pellets
	Reserve	3,640,008	Pellets
	U.S. Steel	618,355	Sinter, Nodules
	Cleveland-Cliffs	476,876	Pellets
1960	Erie	5,733,845	Pellets
	Reserve	4,941,322	Pellets
	U.S. Steel	689,438	Sinter, Nodules
	Cleveland-Cliffs	756,512	Pellets
1961	Erie	7,590,837	Pellets
	Reserve	5,959,307	Pellets
	U.S. Steel	822,013	Sinter, Nodules
	Cleveland-Cliffs	1,580,669	Pellets
1962	Erie	7,710,226	Pellets
	Reserve	5,568,058	Pellets
	U.S. Steel	700,994	Sinter, Nodules
	Cleveland-Cliffs	2,221,277	Pellets
	Hanna	426,027	Pellets
1963	Erie	8,234,815	Pellets
	Reserve	7,697,984	Pellets
	U.S. Steel	792,209	Sinter, Nodules
	Cleveland-Cliffs	3,556,493	Pellets
	Hanna	808,042	Pellets
	Total	98,343,559	

*Mesabi Iron Company figures were compiled from reports in the files of the Minnesota Historical Society. Those for others firms were assembled by Miss Mildred R. Alm of the Mines Experiment Station, largely on the basis of figures reported in *Minnesota Mining Directory* for the years covered. The Cleveland-Cliffs and Hanna plants are in Michigan; all others are in Minnesota.

Footnotes

INTRODUCTION

[1] E. W. Davis, "Taconite: The Derivation of the Name," in *Minnesota History,* 282 (Autumn, 1953). The article concludes that the word "taconite" came "from the Indian name of a mountain or a range of mountains on the western border of Massachusetts." For additional information on various types of ore, see Appendix 1.

[2] In general, the smelting of iron ore involves the removal of oxygen, but after this has been accomplished, many delicate processing steps are required to produce various kinds of iron and steel for modern use. For a relatively nontechnical discussion, see *The Making, Shaping and Treating of Steel* (Pittsburgh, 1957), published by the United States Steel Corporation. On the history of iron manufacture and the early development of smelting furnaces, see Douglas A. Fisher, *The Epic of Steel,* 5–20 (New York, 1963). An important earlier work on this subject is James M. Swank, *Iron in All Ages* (Philadelphia, 1892).

[3] For a discussion of this plant, often called "the birthplace of the American iron and steel industry," see Fisher, *Epic of Steel,* 70–73.

[4] Many other improvements have been made since the 1940s, and by 1960 modern blast furnaces thirty-five feet in diameter were producing about 1,500 tons of pig iron per day from standard ores.

CHAPTER 1 — AN IRON MOUNTAIN ON THE MESABI

[1] On the copper development in the Ontonagon area, see Western Historical Company, *History of the Upper Peninsula of Michigan,* 128–135 (Chicago, 1883); Grace Lee Nute, *Lake Superior,* 156–165 (Indianapolis, 1944); James K. Jamison, *This Ontonagon Country: The Story of an American Frontier,* 55–74 (Ontonagon, 1948).

[2] Walter Van Brunt, ed., *Duluth and St. Louis County, Minnesota,* 61, 66 (Chicago and New York, 1921); Edward F. Greve, "The Development of Lake County, Minnesota," 26, a manuscript in the files of the Minnesota Historical Society.

[3] Otto E. Wieland, "Early Beaver Bay and Its Part in the Discovery of Iron," 2, a manuscript in the files of the Minnesota Historical Society; Greve, "Lake County," 43, 46. On the Vermilion gold rush, see William W. Folwell, *A History of Minne-*

sota, 4:4–8 (St. Paul, 1930); Henry H. Eames, *Report of the State Geologist on the Metalliferous Region Bordering on Lake Superior,* 6, 10 (St. Paul, 1866).

[4] Wieland, "Early Beaver Bay," 4; Walter G. Swart, "Work Done on the Eastern Mesabi Range by Peter Mitchell," a paper presented before the St. Louis County Historical Society in 1923 and published in *Skillings' Mining Review,* September 1, 1923, p. 1 (cited hereafter as *Skillings*), and in the *Duluth News-Tribune,* July 12, 1925; Rollin E. Drake, comp., "A Concise History of the East Mesabi Magnetic Taconite," a mimeographed manuscript of which the Minnesota Historical Society has a microfilm copy; William W. Spalding Autobiography, 32, in the files of the St. Louis County Historical Society, Duluth.

[5] Swart, in *Skillings,* September 1, 1923, p. 5; Jamison, "Ontonagon County Families Established up to the Civil War Period," 23, 43, 108, 119, 169, 188, a manuscript in the Michigan State Library, Lansing, of which the Minnesota Historical Society has a microfilm copy; Fred Wieland to Ernest A. Schulze, August 17, 1934, a letter owned by the St. Louis County Historical Society.

[6] "Ontonagon County Families," 124; Swart, in *Skillings,* September 1, 1923, p. 5.

[7] For maps showing this route, see plates 68 and 77 of Geological and Natural History Survey of Minnesota, *Final Report,* Volume 6 (St. Paul, 1901).

[8] See Grace Lee Nute, ed., *Mesabi Pioneer: Reminiscences of Edmund J. Longyear,* 9 (St. Paul, 1951), for a sketch of what was believed in 1890 to be a cross section of the Lake Superior district. We now know that the "good ore" on the Michigan and Wisconsin ranges pinches out and disappears under Lake Superior before reaching the surface in Minnesota, and that on the Mesabi the taconite rests directly on the quartzite and granite. For a brief account of the opening of the Marquette Range, see F. Clever Bald, *Michigan in Four Centuries,* 237–242 (New York, 1954).

[9] Christian Wieland's comments on "Missabay Height" or "Misabay Heighth," as he called it, may be found in his manuscript "Field Notes," 1872, for the survey of Township 60 North, Range 13 West, p. 62, 63, the originals of which are on file in the Minnesota Secretary of State's office in the Capitol, St. Paul. He notes that the southern slope was timbered with "white pine, Norway pine, pitch pine, birch, poplar, tam[a]rack & spruce . . . in abundance. Along the northern slop[e] there are large granit[e] blo[c]ks scattered." It will be observed that "Mesabi" is spelled in several different ways. For an explanation of its origins and development from the Ojibway word meaning "giant," see Warren Upham, *Minnesota Geographic Names,* 503 (*Minnesota Historical Collections,* Volume 17, 1920).

[10] Swart, in *Skillings,* September 1, 1923, p. 1; see also Albert H. Chester, "The Iron Region of Northern Minnesota," in Geological and Natural History Survey of Minnesota, *Eleventh Annual Report,* 157 (Minneapolis, 1884).

[11] See Hal Bridges, *Iron Millionaire: Life of Charlemagne Tower,* 135 (Philadelphia, 1952); Swart, in *Skillings,* September 1, 1923, p. 8.

[12] As present statistics show, the actual tonnage of iron ore in the Tower area was small. Up to its closing in 1962, the Soudan Mine at Vermilion Lake produced about 16,000,000 tons, while the Peter Mitchell Taconite Mine at Babbitt in five years has produced over twice as much ore. Mildred R. Alm, *University of Minnesota Mining Directory, 1963,* 166, 187 (Minneapolis, 1963).

[13] Ramsey Diary, September 20, October 22, 23, 1872, in the possession of the Minnesota Historical Society. See also Ramsey's memorandum at the end of the diary, dated April 7, 1872, giving the chemical analysis of a sample of iron ore which showed a "pure metalic iron" content of 58.74; and Wieland's manuscript "Field Notes" and the original 1872 plats of Township 60 North, Ranges 12 and 13 West, in the Minnesota Secretary of State's office. The surveyor filed his notes for Range 13 on September 25, 1872.

[14] Moreover, not until 1842, when the Webster-Ashburton Treaty was signed with

England, was it settled once and for all that the Minnesota iron region belonged to the United States. See Fremont P. Wirth, *The Discovery and Exploitation of Minnesota Iron Lands,* 28–55 (Cedar Rapids, Iowa, 1937); Bridges, *Iron Millionaire,* 155–157; Samuel T. Dana and others, *Minnesota Lands,* Appendix 5 (Washington, D.C., 1960).

[15] Abstracts and deeds to much of the property are in the files of the Reserve Mining Company at Silver Bay. Some of this acreage is now occupied by that firm's Peter Mitchell Mine.

[16] In addition to the three Duluth men, the incorporators of the railroad company were Culver, Dickens, Spalding, Stone, Markell, Bailey, Howard, Hunter, and Ensign. See Spalding Autobiography, 32; Van Brunt, *Duluth and St. Louis County,* 350; Bridges, *Iron Millionaire,* 173. On the land grant, see United States, *Statutes Statutes at Large,* 9:519, 12:3; Minnesota, *Special Laws,* 1875, p. 286.

[17] In addition to the officers, the other incorporators were Linus Stannard, James Mercer, Henry P. Wieland, and William Harris. In exchange for the land they conveyed to the company, Stannard was issued 13,329 shares in his own name and 933 more as the administrator of Willard's estate; William D. Williams, 13,477; Harris, 7,609; Mercer, 5,455; Spalding, 3,639; Henry Wieland, 3,333; and Ramsey, 2,225. See Mesaba Iron Company, "Minute Book," 1, 2, 4, 9, in the collections of the St. Louis County Historical Society.

[18] See "Minute Book," 10, 123, 135. The date of Willard's "sudden death" is given by Swart in *Skillings,* September 1, 1923, p. 8; that of Wieland appears in Otto E. Wieland, "Short History of the Wielands," 3, a manuscript in the files of the St. Louis County Historical Society.

[19] Bridges, *Iron Millionaire,* 140, 153; *Minnesota Legislative Manual,* 1876, p. 152 (St. Paul, 1876).

[20] Stone's transactions with Tower and Chester's expedition are described by Bridges in *Iron Millionaire,* 135–152, and in Van Brunt, *Duluth and St. Louis County,* 350–355. See also Chester, "Explorations of the Iron Regions of Northern Minnesota during the Years 1875 and 1880," a manuscript in the collections of the St. Louis County Historical Society, and Geological and Natural History Survey, *Eleventh Annual Report,* 156, 157. A road cut by Chester's crew from Mesaba to Birch Lake became known as the Syndicate Trail.

[21] Chester, in Geological and Natural History Survey, *Eleventh Annual Report,* 158. The author has examined and sampled this cliff, which was later named "Cliff Quarry." Near the top it is a little better grade of taconite than average and was later mined for experimental purposes; two or three inches below the top layer it averages about the same as that on the rest of the eastern Mesabi.

[22] Bridges, *Iron Millionaire,* 152, 153, 160.

[23] Bridges, *Iron Millionaire,* 170, 172–186.

[24] Bridges, *Iron Millionaire,* 186, 213–223. See also Folwell, *Minnesota,* 4:16; *Skillings,* December 13, 1924, p. 1, July 25, 1959, p. 4; *Missabe Iron Ranger,* house organ of the Duluth, Missabe and Iron Range Railroad, July, 1959, p. 5. On the land grant, see Minnesota, *Special Laws,* 1885, p. 258.

[25] See Nute, ed., *Mesabi Pioneer,* 7, 9–22, for Edmund Longyear's description of his experiences while prospecting on the eastern Mesabi.

[26] "Minute Book," 73, 136, 137. On the Mountain Iron Mine, see June D. Holmquist and Jean A. Brookins, *Minnesota's Major Historic Sites: A Guide,* 168–172 (St. Paul, 1963).

[27] Information on St. Clair, his associates, and their business activities on the Mesabi may be found in *History of East Mesabi Magnetic Taconite,* Chapter 1, p. 6–13, a mimeographed volume of which the Minnesota Historical Society has a copy. The author was well acquainted with both St. Clair and Williams.

[28] Articles of incorporation for these firms are on file in the Minnesota Secretary of State's office. The incorporators are listed as Williams, St. Clair, and Arthur Howell, who was probably an agent for Samuel Mitchell. See also *History of East Mesabi Magnetic Taconite,* Chapter 1, p. 12.

CHAPTER 2 – THE MINES EXPERIMENT STATION ENCOUNTERS TACONITE

[1] See C. W. Hall, *The University of Minnesota: An Historical Sketch,* 19, 51 (Minneapolis, 1896); James Gray, *The University of Minnesota, 1851–1951,* 123–126 (Minneapolis, 1951). *Ed.*

[2] Dean Appleby retired in 1935 and died in 1941. After his retirement, the School of Mines became a part of the university's Institute of Technology. While this followed the modern trend toward consolidation and is probably for the general good, something is also lost in bigness.

[3] In actual practice it has since been found necessary to crush the taconite much finer, to a powder having substantially no particles larger than a five hundredth of an inch in diameter.

[4] Detailed results of this test are recorded under "Ore Number 199, Test 5," in the files of the Mines Experiment Station at Minneapolis.

[5] A magnetic separator is simply an ore-processing machine that uses magnets to effect a separation of the magnetic constituents of the ore from the nonmagnetic ones. If the machine is designed to work on coarse dry ore, it is usually called a dry cobber. If it works in water on intermediate sized ore particles, it is called a wet cobber. If it works on fine ore in water, it has no definite name but is usually called a finisher magnetic separator. Several designs of all these machines are now made by various manufacturers.

[6] For a more detailed description of the magnetic tube and the method of determining the magnetic iron assay of a sample of ore, see Davis, *Magnetic Concentration of Iron Ore,* 56–62 (University of Minnesota, Mines Experiment Station, *Bulletins* No. 9— Minneapolis, 1921) and Appendix 1, below. This laboratory machine developed by Dr. Davis is still marketed by the Dings Magnetic Separator Company of Milwaukee as the "Davis Magnetic Tube." It is also made by the Phelan Manufacturing Corporation of Minneapolis. *Ed.*

[7] On 1914 ore shipments, see Lake Superior Iron Ore Association, *Lake Superior Iron Ores, 1938,* p. 300 (Cleveland, 1938).

[8] The machine was apparently so named because it was originally made from a log or tree trunk. For a more detailed description, see Davis, *Magnetic Concentration,* 67–71. *Ed.*

[9] Detailed results of these tests may be found under "Ore Number 199, Tests 50, 53," in files of the Mines Experiment Station. Mesh sizes in this book refer to Tyler Standard Screens. The mesh scale reflects the number of square openings per linear inch. Ordinary window screen, for example, is about 14 mesh, that is, it has 14 openings in a linear inch or 196 openings in a square inch, and each opening is 0.046 of an inch in size. The higher the mesh number the finer the screen. Thus 150 mesh has 22,500 openings in a square inch and each opening is 0.0041 of an inch in size. A mesh number of 325 means 105,625 openings per square inch with openings 0.0017 of an inch in size. When a mesh number is preceded by a minus sign, it indicates that the material will pass through a screen of that size. If the number is preceded by a plus sign, the material will remain on a screen of the size indicated. When one says that ore is crushed to

a given mesh size, say to 100, this means that the material is largely -100 mesh but a small amount of it — perhaps 1 or 2 per cent — will be +100 mesh.

[10] See Arnold Hoffman, *Fortieth Anniversary of the Mesabi Iron Company, 1919–1959: A Short History,* 14 (New York, 1959).

[11] Hoffman, *Mesabi Iron Company,* 15, 16.

[12] Woodbridge to MacKelvie, October 19, 1914, a copy of which is in the files of the Minnesota Historical Society. We now know that the average of the property is about 30 per cent total iron, or about 24 per cent magnetic iron, which is the only type recoverable by magnetic concentration. Therefore, the samples tested by Woodbridge and Jordan must have been taken from a high-grade area.

[13] Swart's "Report on the Magnetic Deposits of the East Mesabi Range," dated July, 1915, appears as Chapter 3 in *History of East Mesabi Magnetic Taconite.* Quoted material may be found on page 24.

[14] See William K. Montague, "A Chronology of the Development of Reserve Mining Company," 2, 3. The Minnesota Historical Society has a microfilm copy of this useful manuscript. The later dissolving of the Dunka-Mesaba firm is discussed in *Skillings,* October 13, 1951, p. 17.

[15] A complete file of these reports, consulted throughout the preparation of this study, may be found at the Mines Experiment Station.

[16] See Nute, ed., *Mesabi Pioneer,* 10–12.

[17] The study is entitled *The Magnetite Deposits of the Eastern Mesabi Range* (Minnesota Geological Survey, *Bulletins* No. 17 — Minneapolis, 1919).

[18] The author is indebted to George M. Schwartz, professor emeritus of the geology department, University of Minnesota, for his help in the preparation of this discussion. Many geologists have since studied the Mesabi. For a recent work which summarizes many earlier studies, see David A. White, *The Stratigraphy and Structure of the Mesabi Range, Minnesota* (Minnesota Geological Survey, *Bulletins* No. 38 — Minneapolis, 1954). On the Duluth Gabbro mentioned below, see Frank F. Grout and others, *The Geology of Cook County, Minnesota,* 2 (Minnesota Geological Survey, *Bulletins* No. 39 — Minneapolis, 1959), and Figure 15, below.

[19] See Grout and Broderick, *Magnetite Deposits,* 36; more recent tonnage estimates are five to six times greater. On the term "septaria," see Grout and Broderick, *Magnetite Deposits,* 23. Although only the narrow band described is truly septaria, the word, like taconite, has with usage lost some of its geological meaning. Taconite mine and plant operators now often use septaria to mean any taconite that is difficult to concentrate.

[20] On this plant, see Swart to Jackling, April 7, 1918, Mesabi Iron Company Papers, in the Minnesota Historical Society. The original is in the files of Hayden, Stone and Company, New York City. See also Davis, "Pioneering with Taconite: The Birth of a Minnesota Industry," in *Minnesota History,* 276–279 (Autumn, 1955). *Ed.*

[21] The wet cobber is more fully described in Davis, *Magnetic Concentration,* 65–67. *Ed.*

[22] Up to the present time, no machine has been developed that can, on a commercial scale, duplicate the work of the Magnetic Tube Separator on small samples of taconite ground to 300 mesh or finer.

[23] For an explanation of briquetting, see Chapter 4, page 83. In the nodulizing process, the fine ore is fed into a long, inclined, brick-lined, slowly rotating tube about eight feet in diameter and a hundred feet long. As the ore moves down the inclined tube, it meets burning gases introduced into the lower end of the rotating tube. Thus the temperature of the ore particles is raised until they soften and fuse together into roughly spherical masses of mixed sizes up to six

or more inches in diameter. Nodules are high in bulk porosity but rather low in microporosity.

[24] See *Lake Superior Iron Ores, 1938,* p. 300, for the average analysis of Lake Superior ore in 1918.

[25] Swart, "Summary and Report of Place and Operations of Mesabi Iron Company, Babbitt, Minnesota, July, 1931," p. 68. Copies of this 260-page typewritten report are in the files of the Minnesota Historical Society and the Bankers Trust Company, New York City.

[26] For the names of some visitors, see Davis, "Eastern Mesabi Magnetic Taconite in the Early Days," in *History of East Mesabi Magnetic Taconite,* Chapter 2, p. 28. *Ed.*

[27] Among these were the lands of the Oriental Granite and Iron Company just west of the Williams-St. Clair holdings, now part of the Erie Mining Company property. At Black River Falls, Wisconsin, several high mounds of magnetic iron ore were located, and the Mesabi Syndicate took leases on some of them. These are now controlled by the Inland Steel Company. Options were also taken on what was known as the Penokee Gap property near Mellen on the western part of the Gogebic Range in Wisconsin, and on the Moose Mountain property in Ontario, now owned by The Hanna Mining Company.

[28] These reports have since been scattered. The author consulted a number of them in the files of Reserve Mining Company at Silver Bay.

[29] See Hoffman, *Mesabi Iron Company,* 19.

CHAPTER 3 – THE FIRST COMMERCIAL TACONITE PLANT

[1] The actual cost of the completed plant was $3,849,613, but the entire project, including lands, development, and the Duluth experimental plant totaled $6,835,525. See Swart, "Summary," 4; Jackling to Hayden, July 22, 1920, Mesabi Iron Company Papers, Minnesota Historical Society.

[2] See "Mesabi Iron Company Formed," in *Iron Age,* January 15, 1920, p. 197. On the firm's later involved financial history, see *History of East Mesabi Magnetic Taconite,* Chapter 1, p. 16; Hoffman, *Mesabi Iron Company,* 19–22. Note that the original Mesaba Iron Company organized by the Ontonagon Pool is spelled with an "a" like the town of Mesaba, while the later company's name is spelled, like the Mesabi Range, with an "i."

[3] Swart, "Summary," 96. It amounted to 1,706 tons and assayed 62.26 per cent dry iron and 11.72 per cent silica.

[4] On the selection of mine and plant sites, see Jordan, Davis, and Counselman, "A Reconnaissance Trip to Eastern Mesabi, May 6–8, 1917," in *History of East Mesabi Magnetic Taconite,* Chapter 4. *Ed.*

[5] No summer cabins or resorts existed on Birch Lake in those days, and the whole valley between the lake and the Giant's Range was available for tailings. There was one ranch belonging in 1920 to Robert Scott which might eventually have been affected if tailings from all the Williams-St. Clair taconite lands were deposited in this valley. The old Scott property was purchased in 1948 by the Reserve Mining Company and is the site of present-day Babbitt.

[6] For more detailed data on the drilling operation and a discussion of the entire project, see Arthur B. Parsons, *Operations of Mesabi Iron Co.,* a reprint from issues of the *Engineering and Mining Journal-Press,* for January 26 and February 2, 1924.

[7] See Figure 3, above, for the Duluth Flow Sheet.

[8] Birch Lake, with an elevation of 1,425 feet, was the principal source of water. It was also the most expensive to pump from, for it was two and a half miles away and three hundred feet below the elevation of the reservoir. The two smaller lakes were closer and slightly above the elevation of the reservoir.

[9] See Davis, in *Minnesota History,* 279–283 (Autumn, 1955), for additional detail on the construction of the plant. A long article entitled "Magnetic Ore Treatment Plant at Babbitt," describing with some inaccuracies the camp and town, appears in *Skillings,* July 31, 1920, p. 1. See also Davis, in *History of East Mesabi Magnetic Taconite,* Chapter 2, p. 33–35; Van Brunt, *Duluth and St. Louis County,* 666. *Ed.*

[10] Judge Babbitt died on February 20, 1920. See *Who Was Who in America,* 40 (Chicago, 1942); Van Brunt, *Duluth and St. Louis County,* 666. *Ed.*

[11] Dwellings and other structures of the Mesabi Iron Company are shown on a map of Babbitt, dated June 5, 1924, which was prepared for the firm by Johnson and Higgins of New York. The Minnesota Historical Society has a copy. Many of these houses have since been sold to Babbitt residents and have been moved to Birch Lake, where they are used as summer cottages.

[12] Van Brunt, *Duluth and St. Louis County,* 667; Swart, "Summary," 25, 257; attendance records for District 83 in the Minnesota State Archives, St. Paul.

[13] See "Babbitt, Minnesota: An Historical Review," in *Skillings,* February 6, 1960, p. 5.

[14] *Mesabi Sinter,* a leaflet in the Minnesota Historical Society's files. The advertised analysis of other ingredients was 9 per cent silica, 0.027 manganese, 0.20 phosphorus, 0.67 alumina, 0.10 lime, 0.50 magnesia, 0.005 sulphur, 0.03 titanium oxide, no moisture.

[15] Swart, in *History of East Mesabi Magnetic Taconite,* Chapter 3, p. 16, 23.

[16] The Lake Erie method of pricing ores has been attacked and tested in the courts, but has always been sustained. It is still in use, with modifications, and is the basis on which the state of Minnesota computes tax values. For a fuller discussion, see Appendix 1.

[17] Before 1903 Old Range ore received premiums as high as a dollar per ton, but since that time they have averaged about 25 cents. See *Lake Superior Iron Ores,* 288 (revised, 1952). Swart's 1915 calculations for the proposed Mesabi Sinter may be found in *History of East Mesabi Magnetic Taconite,* Chapter 3, p. 14.

[18] For the 1921 figures, see *Lake Superior Iron Ores, 1938,* p. 300. Swart's report to Jackling in 1915 (*History of East Mesabi Magnetic Taconite,* Chapter 3) contains no discussion of the silica analysis of the proposed concentrate. It is known, however, that a 60 per cent iron sinter will contain about 14.25 per cent silica; a 61 per cent sinter about 12.85 per cent; and a 62 per cent sinter about 11.45 per cent, etc. For a more complete discussion of the subject, see T. T. Read, P. H. Royster, and T. L. Joseph, *Effect of Silica in Iron Ore on Cost of Pig-Iron Production,* (U. S. Bureau of Mines, *Report Investigation,* No. 2560 — Washington, D.C., 1924); and Appendix 1, below.

[19] Swart, "Summary," 229; interview of the author with Clement K. Quinn, 1962.

[20] For expenditures and other data on the building of the plant, see Dillon to Mesabi Iron Company stockholders, December 4, 1920, March 7, 1921, December 15, 1921, Mesabi Iron Company Papers; *Skillings,* August 25, 1923, p. 16–18. *Ed.*

[21] See Appendix 3, below, for a record of shipments from the Babbitt plant.

[22] Letters in the possession of C. K. Quinn, Duluth. The Minnesota Historical Society has copies of those quoted.

[23] Actually the flow sheet was developed at the Mines Experiment Station. See "Ore Number 655," Mines Experiment Station, April 19, 1924, for the results of tests using it.

[24] Swart, "Summary," 126, 128, 200. Since Mesabi was a stock company, Swart's figures do not include interest on investment or amortization.

[25] See Jackling to the board of directors, June 23, 1923, May 8, 1924, in Mesabi Iron Company Papers.

[26] The Ford rumor was reported in the *Duluth News-Tribune,* March 10, 11, 1923. *Ed.*

[27] Swart, "Summary," 115, 232; *Skillings,* June 28, 1924, p. 7. The last shipment from this stock pile was made in 1929; in 1931 an estimated 5,788 tons of fine sinter remained on hand.

[28] See *Lake Superior Iron Ores, 1952,* p. 288.

[29] See *Lake Superior Iron Ores, 1952,* p. 275.

[30] See *Lake Superior Iron Ores, 1952,* p. 277. Before the shutdown in 1924, Mesabi Iron Company stock was being quoted as high as $13.50 a share. It never ceased entirely to be quoted on the New York Curb Market, but after the shutdown, the price dropped gradually to about 25 cents a share. It was the first and only taconite stock to be offered for sale to the public up to 1964. See Hoffman, *Mesabi Iron Company,* 22–24; *History of East Mesabi Magnetic Taconite,* Chapter 1, p. 16.

CHAPTER 4 – BACK TO THE LABORATORY

[1] Among those consulted at the university were: Regents Fred Snyder, John Williams, and Pierce Butler, President Lotus Coffman, Albert J. Lobb, comptroller, Professors William H. Emmons of the department of geology, William H. Hunter of chemistry, and Levi B. Pease, Elting H. Comstock, and Peter Christianson of the School of Mines. Others with whom we talked included Governor J. A. O. Preus, Edmund J. Longyear, and Russell M. Bennett, mining explorers and engineers, and C. K. Leith, the well-known authority on the geology of the Mesabi Range.

[2] On the new building, see Gray, *The University of Minnesota,* 252; Appleby, "Minnesota Mines Experiment Station," in *Minnesota Techno-log,* 180–183 (March, 1926). *Ed.*

[3] Semitaconite is the name chosen for legislative purposes to distinguish the nonmagnetic taconite of the western Mesabi from the magnetic taconite at the eastern end of that range. In the former the iron is largely in the form of hematite; in the latter it is magnetite. The Cooley plant operated during the 1934 and 1935 shipping seasons as an experimental unit under the direction of the station's staff. The university's interest in it was then sold to Butler Brothers, who put the unit into commercial operation from 1936 through 1938. The process was successful technically but did not work out commercially, because the fuels available at that time made the cost too high. See Appendix 2, below, and Davis, *Magnetic Roasting of Iron Ore* (University of Minnesota, Mines Experiment Station, *Bulletins* No. 13 — Minneapolis, 1937) for a description of the plant and the process. See also *Minneapolis Journal,* May 22, 1934. Interest in magnetic roasting has been revived, and the process is again under investigation by The Hanna Mining Company, the Oliver Iron Mining Division of United States Steel Corporation, the United States Bureau of Mines, the Mines Experiment Station, and others. For information on the manganese and iron powder projects, see Firth, *Iron Powder* (University of Minnesota, Mines Experiment Station, *Information Circulars* No. 3 — Minneapolis, 1943); *Developing Natural and Human Resources in Minnesota,* 20 (Minnesota Iron Range Resources and Rehabilitation Commission, *Reports,* 1950–52 — St. Paul, 1952); *St. Paul Dispatch,* September 10, 1940; *Minneapolis Star Journal,* September 28, 1940, March 12, 13, 26,

December 27, 1942; *Minneapolis Times,* November 3, 1944; *Minneapolis Star,* October 16, 1951.

[4] Among them were: Walter L. Maxson, an expert on fine grinding then employed by the Allis-Chalmers Manufacturing Company of Milwaukee and later director of research for the Oliver Iron Mining Division of United States Steel Corporation; Ted Counselman, whom we had known at Babbitt and who was then on the engineering staff of the Dorr Company of New York; Henry K. Martin, an expert on classification with the Colorado Iron Works of Denver and later an engineer with Oglebay Norton and Company of Cleveland; R. W. Whitney of Butler Brothers, with whom we discussed drilling and mining methods for the hard taconite; and R. A. Manegold of the Dings Magnetic Separator Company of Milwaukee, to name a few.

[5] Especially interesting were discussions in the late 1930s and early 1940s with such men as William A. Haven of Arthur G. McKee and Company of Cleveland, George C. Hewitt of Wheeling Steel Corporation, Frank E. Vigor and Leo F. Reinartz of Armco Steel Corporation, S. O. Hobart of the War Production Board, and Earle Smith and Earl M. Richards of Republic Steel Corporation.

[6] For explanations of microporosity and the changing of magnetite to hematite, see Appendix 2, below.

[7] Reserve was organized as a Minnesota corporation on March 24, 1939, by Oglebay Norton as agents for American Rolling Mill Company (now Armco), Wheeling Steel Corporation, Montreal Mining Company, and Cleveland-Cliffs Iron Company. On this point, see *Skillings,* March 20, 1937, p. 10, January 29, 1938, p. 5, October 28, 1939, p. 1. Mesabi stockholders approved the lease on June 28, 1939, and Reserve exercised its option and took over the property on October 24, 1939. The lease provided that if and when Reserve operated the property any profit would be divided, one-third to Mesabi and two-thirds to Reserve. Reserve also agreed to pay the older company's debts in the amount of $268,655. See Jackling to stockholders, May 20, October 24, 1939, in Mesabi Iron Company Papers, Minnesota Historical Society. See also *Skillings,* May 20, 1939, p. 7, July 1, 1939, p. 3, and October 28, 1939, p. 1, for the terms of the option, and Chapter 6, below. *Ed.*

[8] See also Wade to Johnston, June 20, July 3, September 3, 1941, in the files of the Mines Experiment Station.

[9] Closed-circuit grinding of taconite is effective with ball mills, but not with rod mills, which we found operated best in open-circuit, or one-pass, grinding. For a technical and somewhat controversial discussion of ball mill grinding, see Davis, "Fine Crushing in Ball-Mills," in American Institute of Mining and Metallurgical Engineers, *Proceedings,* 61:250–294 (New York, 1919).

[10] Babbitt taconite averages about 24 per cent magnetic iron. It can be shown mathematically that for a 64 per cent iron concentrate, if the tailings rejected assay 3 per cent iron, they will represent an iron loss of 8 per cent of the iron in the taconite. If they assay 4 per cent iron, the loss will be 10 per cent; at 5 per cent, the loss will be 14 per cent; and if they assay 10 per cent iron, the loss will be 30 per cent. We decided to try for about 4 per cent magnetic iron tailings, which would be about a 10 per cent loss, or a 90 per cent recovery of the iron. See Appendix 1, below.

[11] For an exhaustive discussion of the microstructure of magnetic taconite, see James N. Gundersen and George M. Schwartz, *The Geology of the Metamorphosed Biwabik Iron-Formation, Eastern Mesabi District, Minnesota* (Minnesota Geological Survey, *Bulletins* No. 43 — Minneapolis, 1962).

[12] Commercial grinding here means all particles finer than 300 mesh. However, this limitation is changing. In the Empire plant, opened by Cleveland-Cliffs in 1963 on the Marquette Range in Michigan, all particles of concentrate are ground finer than 400 mesh. The early samples sent to the Mines Experiment Station

from Babbitt were taken from locations near Peter Mitchell's first test pit and contained few dusty middlings.

[13] See Schwartz, *Iron-ore Sinter* (American Institute of Mining and Metallurgical Engineers, *Technical Publications* No. 227 — New York, 1929).

[14] The results of the work discussed here and below are recorded in "Reverberatory Smelting Process for Iron Ores," a mimeographed report dated July 1, 1931, and in seventeen bound volumes covering the years from 1927 to 1931, all of which are in the files of the Mines Experiment Station. See also Davis, "Test Pavement Laid with Cast Iron Road Plates," in *Steel,* November 26, 1934, p. 30; "Cast Iron Test Pavement Laid," in *The Foundry,* November, 1934, p. 32, 77; "Cast Iron Paving Blocks," in *Mining Congress Journal,* November, 1936, p. 40–42, 47; *Minnesota Chats,* October 22, 1934; *Minnesota Daily,* February 26, 1936; *Eveleth News,* October 26, 1939; a file of miscellaneous clippings, 1936–39, in Mines Experiment Station's Scrapbooks. On direct reduction, see Appendix 2, below. *Ed.*

[15] See N. V. Hansell, "The Briquetting of Iron Ores," in American Institute of Mining Engineers, *Transactions,* 43:394–411 (New York, 1912).

[16] The work of the Mines Experiment Station using this process is described by Davis and Wade in *Agglomeration of Iron Ore by the Pelletizing Process* (Mines Experiment Station, *Information Circulars* No. 6 — Minneapolis, 1951). *Ed.*

[17] Greenawalt and Stehli developed what is known as the "Greenawalt Sintering Process" for agglomerating fine iron ore and flue dust. It is still used by the Bethlehem Steel Corporation at its Lebanon, Pennsylvania, plant. It is an intermittent operation that has been largely replaced by the continuous Dwight and Lloyd process. See Albert F. Plock, "The Reclamation of Flue Dust for Furnace Use," in *Iron Trade Review,* May 1, 1913, p. 1017.

[18] The oxidation of magnetite to hematite, like the oxidation of coal, gives off heat. This heat, liberated within the pellets, is sufficient to raise the temperature from 1900 degrees to the 2300 degrees required for pelletizing.

[19] We made no attempt to recover the heat in the red-hot pellets discharged from the furnace. Shaft furnaces now in use in modern pelletizing plants are equipped with heat recuperators, and the fuel consumption is only about one-fourth of that required by the original Mines Experiment Station furnace, which had no recuperators.

[20] Cooke's findings were published under the title "Microstructures in Iron-Ore Pellets" in American Institute of Mining and Metallurgical Engineers, *Transactions,* 193:1053–58 (1952). See also Chapter 11, note 14, below.

[21] Firth's paper, entitled "Agglomeration of Fine Iron Ores," appears in American Institute of Mining and Metallurgical Engineers, Iron and Steel Division, Blast Furnace and Raw Materials Committee, *Proceedings,* 4:46–69 (1944). Additional information on pelletizing may be found in Davis and Wade, *Agglomeration of Iron Ore.*

CHAPTER 5 — TAXES AND TACONITE IN MINNESOTA

[1] The results are filed under "Ore Number 916" in the Mines Experiment Station's files. The rock of this district resembles taconite only in that the good iron ore it contains is in the magnetic state.

[2] These taxes on iron ore are covered by an amendment to Article 9, Section 1, of the Minnesota Constitution approved by the voters in 1922. See Minnesota, *Session Laws,* 1921, p. 1000–1002 for the text of the proposed amendment and Minnesota, *General Statutes,* 1923, p. 2373. For a discussion of iron ore taxation at the time covered here and below, see Minnesota Interim Commission on Iron Ore Taxation, *Report,* 4–40 (1941). For a more recent discussion, see James B.

McComb, *Iron Mining and Taxes in Minnesota* (Macalester College, Bureau of Economic Studies, *Bulletins* No. 13 — St. Paul, 1961, revised, 1963).

[3] For information on such a proposal made by the author and the reaction to it, see *Duluth Herald,* March 13, 1940; *Minneapolis Star Journal,* March 14, 24, 1940; *St. Paul Pioneer Press,* March 17, 1940; *Eveleth News,* March 23, 1940; *Virginia Daily Enterprise,* March 22, 1940. *Ed.*

[4] For fuller recent discussions of this complex situation, see Hyman Berman, "Education for Work and Labor Solidarity: The Immigrant Miners and Radicalism on the Mesabi Range," and Timothy L. Smith, "Factors Affecting the Social Development of Iron Range Communities," two papers presented at a Conference on the Role of Education in the Minnesota Iron Range Towns, sponsored by the University of Minnesota and the Fund for the Advancement of Education, on October 18 and 19, 1963. Mimeographed copies are in the collections of the Minnesota Historical Society.

[5] On Power, see also Van Brunt, *Duluth and St. Louis County,* 560, 971.

[6] For one such account of Hibbing, see Nathan Cohen, "Razzle-Dazzle Village," in *American Mercury,* 346–350 (March, 1944), and a Hibbing resident's answer in the same volume, page 760 (June, 1944). Cohen's article was reprinted in the *Reader's Digest,* 101–103 (May, 1944).

[7] The earliest legislative attempt to limit local range taxation was in 1921. See *Session Laws,* 1921, p. 646. For quoted material, see Interim Commission, *Report,* 101, 103, 119 (1941).

[8] See *Range Facts,* July 2, 4, 1940; *Queen City Sun* (Virginia), July 5, 1940. Charts used in this talk appear in Davis, *The Iron Ore Deposits of Minnesota: The Effect of Existing Tax Laws on the Utilization of This Great Natural Resource* (University of Minnesota, *Bulletins,* Volume XL, No. 27 — Minneapolis, 1937). For other speeches by Dr. Davis in this period, see *Duluth Herald,* December 16, 1940; *Minneapolis Star Journal,* December 17, 1940; *Chisholm Tribune-Herald,* December 19, 1940; *Virginia Daily Enterprise,* January 31, 1941; *Range Facts,* February 6, 1941. *Ed.*

[9] The talk was reported in *Virginia Daily Enterprise,* December 3, 1940; *Range Facts,* December 5, 19, 1940. The figures on plant costs were estimated on the basis of prices in 1940; in 1960 the actual cost was nearly twice this amount.

[10] The definitions read as follows: "For the purpose of this law, 'taconite' is defined as ferruginous chert or ferruginous slate in the form of compact, siliceous rock, in which the iron oxide is so finely disseminated that substantially all of the iron-bearing particles of merchantable grade are smaller than 20 mesh. Taconite may be further defined as ore-bearing rock which is not merchantable as iron ore in its natural state, and which cannot be made merchantable by simple methods of beneficiation involving only crushing, screening, washing, jigging, drying or any combination thereof." See *Session Laws,* 1941, p. 692.

[11] See *Session Laws,* 1941, p. 1075–1077.

[12] See *Session Laws,* 1941, p. 1077–1079.

[13] For the progress of the bills (House File No. 1292 and Senate File No. 1171) through the legislature discussed here and below, see the *House Journal,* 1941, p. 843, 1292, 1295, 1475, 1680, 1713, 1715, 1741, 1977; *Senate Journal,* 1941, p. 709, 768, 907, 913, 1309, 1349. *Ed.*

[14] For its text see *Session Laws,* 1941, p. 692–694. The Minnesota legislature has passed a substantial body of subsequent legislation dealing specifically with taconite. The author is indebted to Walter N. Trenerry of St. Paul for the following summary of significant laws. See *Session Laws,* 1945, p. 456, 473; 1947, p. 347, 510; 1951, p. 675, 1028; 1955, p. 948, 1130–1136, 1562–1566; 1957, p. 438, 711–713, 1070; 1959, p. 1264, 1666–1672; 1961, p. 707, 1609; 1963, p. 129, 156, 327, 913, 1700.

[15] In the Organic Act of March 3, 1849, creating Minnesota Territory, Congress

reserved sections 16 and 36 in each township for the support of the schools. Additional land grants (including some for the support of a state university) were provided for in the Enabling Act of February 26, 1857, creating the state of Minnesota. The income from these lands goes into permanent state trust funds. Some of the land allotted in northern Minnesota was later found to contain iron ore, and mining companies pay royalties and rentals for its use. Before the taconite tax law was passed in 1941, a number of parcels had never produced any income because they contained only taconite. From 1941 to 1963 about $4,-805,175 was collected by the state in rents and royalties from taconite lands. Of this about $1,300,000 came from lands allotted to the public schools, and $2,300,-000 from those allotted to the university. An additional $1,200,000 came from tax-forfeited lands, and of this 80 per cent was returned to the taxing districts in which the lands were located. Ray D. Nolan, director, Minnesota Division of Lands and Minerals, to the author, October 11, 1963. See also Dana and others, *Minnesota Lands,* 91.

[16] *Session Laws,* 1941, p. 1080–1092. The taconite royalties on state lands have varied upward from a specified minimum of 14 cents per ton. The highest to date, a fifty-year lease of 360 acres of state land to Erie Mining Company signed in April, 1964, calls for a minimum royalty of 27 cents per ton of ore or 81 cents per ton of concentrate, and includes an escalator clause which is tied to the commodity price index. On this lease, see *Minneapolis Star,* April 27, 1964; *St. Paul Dispatch,* April 27, 1964. The basic act for leasing Minnesota mineral lands may be found in *General Laws,* 1889, p. 68–73. Earlier leasing laws mentioning beneficiation, magnetic separation, and low-grade magnetite deposits but not taconite had been passed in 1921 and 1927. See *Session Laws,* 1921, p. 630–640; 1927, p. 525–535. Subsequent laws affecting taconite leases are in *Session Laws,* 1943, p. 329–331; 1947, p. 136; 1949, p. 1068; 1951, p. 1033–1048; 1957, p. 968.

CHAPTER 6 – TACONITE COMES TO LIFE AGAIN

[1] Pickands Mather is an operating and management organization producing iron ore for various steel companies. It is customary for such companies to organize separate firms for the operation of individual mines or projects. The Erie Mining Company, composed of Bethlehem Steel Corporation, Youngstown Sheet and Tube Company, Interlake Iron Corporation, and the Steel Company of Canada, Limited, is one of a number of such firms currently managed by Pickands Mather. See R. H. Ramsey, "Teamwork on Taconite: The Story of Erie Mining Co.'s Commercial Taconite Project," in *Engineering and Mining Journal,* March, 1955, p. 86.

[2] On Erie's leases, see *Range Facts,* June 12, 1941; *Chisholm Tribune-Herald,* June 12, 1941; *St. Louis County Independent* (Hibbing), October 31, 1941. *Ed.*

[3] On the work of the Hibbing laboratory, see Ramsey, in *Engineering and Mining Journal,* March, 1955, p. 89–91; *Skillings,* October 9, 1943, p. 1, November 6, 1954, p. 2. See also F. D. DeVaney, "Flotation of Lake Superior Ores," in *Skillings,* February 11, 1950, p. 1, 4, 6, 15.

[4] See Chapter 4, note 7.

[5] On this matter and Dr. Davis' role in it, see his *Lake Superior Iron Ores and the War Emergency: A Report Presented to Materials Division of the War Production Board,* May 20, 1942, a mimeographed copy of which is in the possession of the Minnesota Historical Society; *Range Facts,* July 30, December 24, 1942; *St. Paul Dispatch,* November 17, 19, 1942; *Cleveland Plain Dealer,* November 25, 1942; *Minneapolis Tribune,* December 15, 16, 17, 1942. The Leith works are: *The Mesabi Iron-bearing District of Minnesota* (U. S. Geological Survey, *Monographs* No. 43

— Washington, 1903), and in collaboration with Charles R. Van Hise, *The Geology of the Lake Superior Region* (U.S. Geological Survey, *Monographs* No. 52 — Washington, 1911). *Ed.*

[6] On Oliver's plans, see *Range Facts,* December 9, 1943. The firm had pioneered iron ore concentration on the Mesabi by building the range's first washing plant at Coleraine in 1907. For a description of it, see Charles E. Van Barneveld, *Iron Mining in Minnesota,* 179–183 (University of Minnesota, Mines Experiment Station, *Bulletins* No. 1 — Minneapolis, 1913).

[7] For a few of the many articles which appeared, see Warner Olivier, in the *Saturday Evening Post,* November 14, 1942, p. 22, 121; *Fortune,* November, 1945, p. 128–139, 259, 260, 262, 264; *Business Week,* May 11, 1946, p. 19–21. The influential report of the Special Committee Investigating the National Defense Program, the so-called "Truman Committee," also appeared during this period. For the group's comments on taconite, see 78 Congress, 1 session, "Interim Report on Steel," 11, in *Investigation of National Defense Program, Senate Reports* No. 10, Part 3 — Washington, 1943.

[8] Straub, "Report on Transportation of Iron Ore Concentrate from the Mesaba [*sic*] Range to Lake Superior," a typewritten document dated September, 1942, in the files of the Mines Experiment Station. The Minnesota Historical Society has a copy.

[9] In 1945 estimates indicated a total operating cost of $5.82 per ton of product delivered to lower lake ports. At that time, the Lake Erie price calculation for Old Range Bessemer ore assaying 65 per cent iron gave a value of $5.98 per ton, thus indicating a profit of 16 cents a ton. However, this was straight operating cost with no charge for amortization or interest on the investment. See Davis, "Beneficiation of Eastern Mesabi Magnetic Taconite," in Mines Experiment Station, *Information Circulars* No. 5, p. 34 (Minneapolis, 1945). See also Richard F. Pulver, "Electric Power and the Iron Mining Industry," 14–21 (University of Minnesota, Center for Continuation Study, *Twelfth Annual Mining Symposium* — Minneapolis, 1951).

[10] For information on fuel costs at this time, see a letter by Dr. Davis in the *Mesabi Daily News,* December 3, 1945. *Ed.*

[11] Data presented here and below is discussed in greater detail in Davis and Wade, *Agglomeration of Iron Ore* (Mines Experiment Station, *Information Circulars* No. 6).

[12] This demonstration was conducted as a part of the 1945 Mining Symposium sponsored by the University of Minnesota. For the papers given there, see Mines Experiment Station, *Information Circulars* No. 5.

[13] Arthur G. McKee and Company and Allis-Chalmers continued to investigate this method of pelletizing or sintering in a pilot plant near Milwaukee. See *Skillings,* April 5, 1952, p. 4.

[14] The technical results are on file at the Mines Experiment Station. See also *Notes on the Pelletizing of Taconite Concentrate on the Traveling Grate at the Mines Experiment Station,* a mimeographed report in the files of the station and the Minnesota Historical Society.

[15] On the building of the plant, see *Skillings,* January 25, 1947, p. 1; May 31, 1947, p. 2; December 20, 1947, p. 8. For a detailed description of its operations, see *Skillings,* October 22, 1949, p. 2; Ramsey, in *Engineering and Mining Journal,* March, 1955, p. 91. The first pellets from this plant (3,255 tons) were loaded by the "Elton Hoyt II" at Two Harbors on June 29, 1950. They were destined for the Steelton plant of Bethlehem Steel. D. M. Chisholm to the author, November 20, 1962.

[16] See Firth, in American Institute of Mining and Metallurgical Engineers, *Proceedings,* 4:46–69 (1944).

[17] For accounts of the Ashland plant, inaccurate in some details, see Morton M. Hunt, "Taconite: Iron Ore Bonanza," in *Steelways,* March, 1951, p. 1–5; Hartzell Spence, "Stubborn Taconite Turns to Iron," in *Nation's Business,* August, 1954. p. 40–42.

[18] Working with this ore, Steffensen had one great advantage that we were slow to recognize — the Cornwall ore is one of the easiest to ball. Just why this is so is something of a mystery. Much still remains to be discovered about the balling characteristics of different materials. It is not simply a question of particle size or size distribution, but probably is related to the nature of the fractured surface of the individual particles, involving such obscure phenomena as wetting and surface tension.

[19] B. J. Larpenteur, "Comments on E. W. Davis' and H. H. Wade's Paper," in *Twelfth Annual Mining Symposium,* 1951, p. 56.

CHAPTER 7 – THE CRUCIAL YEARS

[1] On these tests, see reports by William D. Tretheway, N. F. Schulz, and others in a file on "Reserve Mining — Tailings Disposal" in the possession of the Mines Experiment Station. The results were also visually recorded in a 16-millimeter color motion picture entitled "Taconite," released by Reserve in 1955. See also *Minnesota Daily,* December 8, 1950.

[2] One of these great valleys just offshore from Silver Bay, where Reserve's plant was later built, reaches a depth of 800 feet and is many miles long.

[3] On the application, see *Skillings,* February 15, 1947, p. 1; May 31, 1947, p. 2. Hearings were conducted at Two Harbors on June 5, at St. Paul on June 17 and July 3, at Duluth on July 11, and at St. Paul on July 22, September 4, 30, October 21, and November 4. Reports of the hearings appear on the front pages of the *Duluth News-Tribune* for the following day. The discussion below is based on the complete "Transcript of Proceedings," totaling 550 pages on file in the Minnesota Department of Conservation, Division of Waters, St. Paul. The Minnesota Historical Society has a copy. *Ed.*

[4] Hugh S. Bell, *Stratified Flow in Reservoirs and its Use in Prevention of Silting* (United States Department of Agriculture, *Miscellaneous Publications* No. 491 — 1942). The structure is sometimes referred to as Boulder Dam. *Ed.*

[5] *Duluth News-Tribune,* November 5, 1947.

[6] The quoted material appears on pages 131 and 132 of the "Transcript of Proceedings" for September 4, 1947.

[7] *Duluth News-Tribune,* July 12, 1947.

[8] Material quoted here and below is taken from a mimeographed copy of the Minnesota Water Pollution Control Commission, *Chairman's Report of Hearings,* which includes a copy of the permit, in the files of the Minnesota Historical Society. The original permits and their subsequent amendments are on file in the Minnesota Conservation Department, Division of Waters. The permit has twice been amended to allow Reserve to use additional water. The first amendment in 1956 increased the water allowance from 130,000 gallons per minute to 260,000 gallons per minute when the capacity of the Silver Bay plant was increased from 3,750,000 tons per year to 6,000,000 tons per year. The second increase was to 502,000 gallons per minute in 1960 when the expansion to 9,000,000 tons per year was started. Each time the permit was to be amended another hearing was held in Duluth, and representatives of the Northern Sportsmen Club appeared against it. Doubling the flow of water and tailings into the lake simply extended the area of discoloration on its surface from the original four to five feet from the edge of the delta to eight to ten feet. Wilson's find-

ings from the testimony at the hearings that material discoloration except during storms would extend "probably not more than a mile in any direction" have been substantiated. Both the company and the commission make many tests for turbidity, and no "material discoloration" has been detected around the permit area either at the surface or at a depth of twenty feet.

[9] For some typical comments, see "Drama on the Iron Range," in *U.S. Steel News,* January, 1948, p. 10-15; *New York Times,* April 18, June 27, 1948; 80 Congress, 2 Session, *Congressional Record,* part 11, p. A3482-A3484, A3758, A4091; *Hibbing Daily Tribune,* February 22, 1950; "The Taconites Are Ready," an editorial in *Mining Engineering,* September, 1950, p. 933; a large advertisement which appeared in the *Minneapolis Tribune,* October 15, 1950, as well as in many other Minnesota papers; and Rudolph T. Elstad, *Are We at the Crossroads?,* a talk given by the president of Oliver before the Hibbing Chamber of Commerce, January 24, 1950, and later published by the company as a pamphlet. The Mines Experiment Station's Scrapbooks contain a copy. Also of interest is a 44-page booklet issued by the Iron Mining Industry of Minnesota under the title *The New Era in Minnesota Iron Mining* (Duluth, 1951). The title of the *Harper's* article mentioned below is "Steel: The Great Retreat." *Ed.*

[10] The men sent to the university by Armco were: Kenneth C. McCutcheon from management; W. Edward Marshall and Hugh C. Barnes, engineers; James L. Brady as blast furnace superintendent; and six furnace operators: Alexander Boliske, Charles Lake, Frank Rose, and Francis E. Sanders from Armco, and Jack Selway and Charles Williams from the Wheeling Steel Corporation (Wheeling at the time was a part of the Reserve organization). August F. Torreano of Oglebay Norton was in charge of finances and of the employment of miscellaneous labor, largely university students.

[11] For published material on this point, see *Minneapolis Sunday Tribune,* February 22, 1948; W. E. Marshall, "Taconite Pellets in the Blast Furnace," in *Journal of Metals,* April, 1961, p. 308–313. Complete notes on the test are in the confidential files of the Mines Experiment Station. *Ed.*

[12] *Minneapolis Star,* May 28, 31, 1948; *New York Times,* May 29, 1948; *Range Facts,* July 1, 1948. See also, "Correspondence on Taconite Pot," in the files of the Mines Experiment Station for the names of those to whom these replicas were given. The Minnesota, St. Louis County, Lynn, and Ontonagon historical societies each have one. *Ed.*

[13] *Skillings,* September 23, 1950, p. 1; Montague, "Chronology," 12.

[14] *Skillings,* March 24, 1951, p. 2; *Duluth Herald,* September 21, 1951. As first proposed the plant was to have a capacity of 2,500,000 tons, but within a few months this was increased to 3,750,000 tons. See *Skillings,* March 8, 1952, p. 8.

CHAPTER 8 – THE NEXT STEP

[1] *This Is Pilotac,* an undated ten-page booklet issued by the company.

[2] Pilotac opened officially on June 9, 1953, and loaded its first ore car that day. The first full boatload of sinter from the Extaca plant left Duluth on May 3. 1954. On these plants, see *Skillings,* March 11, 1950, p. 2, May 19, 1951, p. 4, December 8, 1951, p. 1, April 18, 1953, p. 1, June 20, 1953, p. 1, May 8, 1954, p. 2; *Hibbing Daily Tribune,* January 25, 1950; *Range Facts,* May 18, 1951; *Minneapolis Star,* June 9, 1953, September 15, 1955; *Duluth Herald,* February 28, 1954. Although Oliver in 1953 announced plans for a commercial plant to be built near Mountain Iron in 1959, these have not materialized to date, and Pilotac is in 1964 the only taconite plant operated by this firm in Minnesota. See *St. Paul Pioneer Press,* August 9, 1953. *Ed.*

[3] See *Minneapolis Star,* October 13, 1949; *Range Facts,* February 24, 1950; *Duluth News-Tribune,* March 15, 1950. The permits are on file in the Minnesota Division of Waters.

[4] For the announcement, see *Duluth News-Tribune,* February 17, 1952. On the later construction of the town of Hoyt Lakes, see *Range Facts,* November 24, 1954; *Duluth Herald,* January 22, 1955; Ramsey, in *Engineering and Mining Journal,* March, 1955, p. 93; *Duluth News-Tribune,* March 13, 1955; *Pioneer Magazine* of the *St. Paul Sunday Pioneer Press,* October 30, 1955; *Minneapolis Tribune,* January 8, 1956; *Erie Mining Company Offers Challenging Careers in Taconite Development,* 13–16, a booklet published by the company in 1957; and an undated mimeographed pamphlet entitled *Welcome to Erie.* While the town and plant were being built in 1954–55, a large trailer camp called Evergreen Park came into being to house the construction workers. An illustrated article in *Picture Magazine* of the *Minneapolis Sunday Tribune* of June 26, 1955, reported that over 600 trailers were parked there. Construction of the town of Hoyt Lakes was started in the spring of 1954 when the site was cleared and graded. The first basement was excavated on May 24, and by the end of the year about 200 houses had been built and about 100 of them were occupied. Hoyt Lakes, like Reserve's towns, was managed by John W. Galbreath and Company of Columbus, Ohio. It was planned with curving streets and varied two, three, and four bedroom houses. Its post office opened early in 1955, and a modern school was completed in 1956. The *Minneapolis Star* of November 23, 1960, reported that Hoyt Lakes had a population of about 3,200. *Ed.*

[5] On the financing, planning, and construction of Erie's plant and harbor discussed here and below, see *Skillings,* February 23, 1952, p. 1, 2; March 1, 1952, p. 1; on Anaconda, November 6, 1954, p. 1, 2, 4, 11; February 5, 1955, p. 1, 2, 3; September 3, 1955, p. 2; October 1, 1955, p. 6; November 26, 1955, p. 2; June 15, 1957, p. 6–9; July 7, 1956, p. 2; July 21, 1956, p. 2; January 5, 1957, p. 22. See also *Duluth News-Tribune,* March 15, 1950, February 17, 1952, December 4, 1953, August 21, 1955; Ramsey, in *Engineering and Mining Journal,* March, 1955, p. 72–93; *Duluth Herald,* August 21, December 4, 1953, February 28, 1954, January 27, 1955; *New York Times,* June 12, 1955; *Range Facts,* November 12, 1953, March 4, 1954. *Ed.*

[6] The Hoyt Lakes mine loaded its first car of taconite on August 14, 1957; the first pellets were produced on September 26, 1957; the first pellet shipment of 10,800 tons from Taconite Harbor was loaded on September 26, 1957, aboard the "J. A. Campbell." These pellets were produced, however, in the Aurora preliminary plant. The first shipment from the Hoyt Lakes plant left Taconite Harbor on October 16, 1957, aboard the "A. B. Wolvin." D. M. Chisholm to the author, November 20, 1962. On the opening of the Hoyt Lakes plant, see *Minneapolis Tribune,* September 28, 1957. After the big plant was in operation, Erie closed its preliminary plant in December, 1954. It had produced over a half million tons of pellets during the six years it was in operation. See *Skillings,* December 7, 1954, p. 4.

[7] The articles appeared respectively in the *Minneapolis Star,* February 12, 1952; *Mining Engineering,* April, 1952, p. 361-363; *Business Week,* October 4, 1952, p. 70-77; *Picture Magazine* of the *Minneapolis Sunday Tribune,* March 15, 1953, p. 5-17, and October 18, 1953, p. 21–28; and *Reader's Digest,* September, 1954, p. 20–24. For others, see Mildred R. Alm, *Bibliography of Taconite, 1920–58* (Mines Experiment Station, 1958).

[8] See *Minneapolis Star,* March 9, 1951. *Ed.*

[9] The author has a copy of the text of this report, *The Activities of the Reserve Mining Company: A Minnesota Corporation,* 109 pages (Minneapolis, June 10, 1952).

[10] Morrill, *Taconite! Sleeping Giant of the Mesabi* (New York, 1944). *Ed.*

[11] The bonds carried an interest rate of 4¾ per cent according to Montague, "Chronology," 13. See also *Range Facts,* January 27, 1952; *Minneapolis Star,* December 26, 1952. The companies were: Metropolitan Life Insurance Company, Equitable Life Assurance Society of the United States, New York Life Insurance Company, Northwestern Mutual Life Insurance Company, Penn Mutual Life Insurance Company, Sun Life Assurance Company of Canada, Massachusetts Mutual Life Insurance Company, Provident Mutual Life Insurance Company of Philadelphia, and Teachers Insurance and Annuity Association of America.

[12] See R. W. Whitney, "Taconite Mining — Past and Future," in Mines Experiment Station, *Information Circulars* No. 5, p. 21–23.

[13] Actually the figures were not so bad. The outsiders did not know, however, that Reserve's power and rail transportation costs were expected to be lower and that the pellets were expected to be worth much more than the computed Lake Erie Base Price indicated. By 1964 pellets were being sold by Cleveland-Cliffs and others at $2.00 above the base price.

CHAPTER 9 — BABBITT REVIVED

[1] On the construction, see Montague, "Chronology," p. 12; "Reserve Mining Co. Starts Taconite Plant at Babbitt" and "Reserve's New Taconite Project," in *Engineering and Mining Journal,* November, 1952, p. 72–79, December, 1956, p. 75–102; *Skillings,* March 24, 1951, p. 2, July 21, 1951, p. 6, October 20, 1951, p. 1, 6, October 25, 1952, p. 1, 2, 8; *Minneapolis Tribune,* September 16, 1951; *Range Facts,* September 27, 1951. *Ed.*

[2] On its name, see *Skillings,* November 17, 1956, p. 4.

[3] During the planning and development stages, August Torreano was in charge of the mine. Later when large-scale mining began, Floyd Erickson became manager. Under his direction a new mining and blasting program was developed. Instead of drilling one line of blastholes behind the working face, fifteen or twenty rows were drilled, loaded with ammonium nitrate, and blasted all at once. Each such blast broke a million tons or more of rock instead of a few thousand. On August 25, 1958, what is said to be the largest open pit blast in history was detonated at the Peter Mitchell Mine, when 794 holes were exploded, breaking loose 1,140,135 tons of taconite — enough to supply the Silver Bay plant for about two weeks. For detail on this blast, see "Sketch & Data of the Largest Open Pit Blast in History," a news release issued by Reserve on August 25, 1958. The Minnesota Historical Society has a copy.

[4] On jet piercing, see D. H. Fleming and J. J. Calaman, "Production Jet-Piercing of Blastholes in Magnetic Taconite," in *Mining Engineering,* July, 1951, p. 585–591; on its use by Reserve, see *Skillings,* March 8, 1952, p. 6, August 29, 1953, p. 6, January 12, 1957, p. 6, January 26, 1957, p. 2, 3, 16, 17.

[5] On the planning and early development of Babbitt, see "Taconite Brings the Modern Mining Town," in *Engineering News-Record,* October 23, 1952, p. 34–39. On the early problems of its residents, see *Duluth News-Tribune,* August 24, 1952; *St. Paul Pioneer Press,* October 26, 1952. *Ed.*

[6] On the opening of the shopping center and the purchase of homes, see *Mesabi Daily News,* January 31, March 17, 1955. On the school, see *Range Facts,* September 9, 1954. For the first marriage in the new town, see the *Ely Miner,* December 2, 1954. The Lutheran Church was the first to be built; see the *Ely Miner,* October 26, November 8, 1956. For stories on later phases of the town's development, see *Duluth Herald,* February 18, 1954, January 20, 1956; *Range Facts,* February 18, September 9, 1954; *Picture Magazine* of the *Minneapolis*

Sunday Tribune, June 19, 1955; *Babbitt News,* June 25, 1958; *Mesabi Daily News,* January 12, 1959, May 16, 1961. Clippings of these and other stories on Babbitt may be found in a file on the new town in the St. Louis County Historical Society at Duluth. *Ed.*

[7] On the election, see *Ely Miner,* October 11, November 1, 1956. For background on the formation of the school district and on Mr. Emanuelson, see *Range Facts,* July 16, 1953; *Duluth Herald,* November 19, 1955; *Ely Miner,* July 25, 1957. On the dedication of the high school, see the *Mesabi Daily News,* October 29, 1960. *Ed.*

[8] On the formation of a businessmen's club, volunteer fire unit, and other organizations, see *Ely Miner,* November 15, 22, 29, 1956. *Ed.*

[9] For additional detail on grinding practices at Babbitt, see E. M. Furness, "Reserve Mining Company: Experience in Crushing and Milling Taconite Ore at E. W. Davis Works, Silver Bay, Minn.," in *Skillings,* February 2, 1957, p. 2, 3, 16. The total steel wear in the Reserve plant in 1964 amounts to about eight pounds per ton of pellets; more than half of this occurs in the rod and ball mills.

[10] At the time the Babbitt plant was being designed, a group of engineers in this country and abroad had developed the theory that most rod mills were being run too slowly and with too large rods. But at least for hard taconite, high-speed, short rod mills loaded with small rods are not the answer.

[11] An 11-page typewritten report, dated January 29, 1952, prepared by Dr. Davis and addressed to R. J. Linney, describes the "Reserve Mining Company Pelletizing Plant" in detail. The Mines Experiment Station has a copy. *Ed.*

[12] Montague, "Chronology," p. 12; *Skillings,* April 25, 1953, p. 1. The first lake shipment went from Two Harbors on May 2, 1953 — 4,918 tons on the "C. W. Gallaway."

[13] See *Skillings,* April 5, 1952, p. 4.

[14] See *Skillings,* October 26, 1957, p. 4.

CHAPTER 10 — SILVER BAY IS BORN

[1] On the contract, see *Range Facts,* September 27, 1951; *Skillings,* October 25, 1952, p. 1, 2, 8. When the construction work was completed, Mr. Lampman joined Mr. Linney's staff. Lawrence J. Molinaro, who went to Silver Bay as manager for a subcontractor, was another who stayed on with Reserve to become superintendent of the important and complex maintenance department.

[2] U.S. Board on Geographic Names, *Decisions List* No. 5701, p. 6 (Washington, D.C., 1957). *Ed.*

[3] Correspondence between Dr. Lakela and Edward Schmid, Jr., of Reserve in November, 1952, is printed in full in *Inspiration,* April, 1953. The Minnesota Historical Society has a file of this magazine, which was published at Duluth, as well as copies of the pertinent correspondence. Dr. Lakela's article appears on pages 265-271 of the Torrey Botanical Club *Bulletin. Ed.*

[4] See *Skillings,* September 13, 1952, p. 10, October 25, 1952, p. 2.

[5] See *Skillings,* March 8, 1952, p. 6.

[6] On the building of Silver Bay discussed here and below, see *Minneapolis Tribune,* December 28, 1952; *Duluth News-Tribune,* August 24, December 23, 1952; *Engineering News-Record,* October 23, 1952, p. 34–39; *St. Paul Pioneer Pioneer Press,* October 26, 1952, November 13, 1955; *Picture Magazine* of the *Minneapolis Sunday Tribune,* June 19, 1955; *Engineering and Mining Journal,* December, 1956, p. 100; and George Grim, in *Minneapolis Tribune,* May 27, 1956. *Ed.*

[7] *Silver Bay News,* October 28, 1958.

[8] On the naming of the town and post office, see *Silver Bay News,* September 20, 1960; *Skillings,* May 8, 1954, p. 2.

[9] See *Silver Bay News,* October 16, December 4, 1956.

[10] On these organizations, see *Picture Magazine* of the *Minneapolis Sunday Tribune,* June 15, 1955; *Minneapolis Star,* September 17, 1956; *Silver Bay News,* May 14, 1957, March 10, 1959. *Ed.*

[11] For a more detailed description, see *Silver Bay News,* February 25, 1958; *St. Paul Pioneer Press,* February 28, 1958.

[12] See *Minneapolis Star,* July 10, 1953.

[13] On the first Mesabi ore shipment, see Paul DeKruif, *Seven Iron Men,* 170 (New York, 1929). The car which Dr. Davis dumped is still in service in 1964 and is now marked with a gold star. On March 17, 1962, it was also the one-millionth car to be dumped into the Silver Bay plant. See *Reserve News Letter,* March 22, 1962. *Ed.*

[14] For a detailed description of the operations in the plant and mine, see *Engineering and Mining Journal,* December, 1956, p. 75–102.

[15] For the start-up of the pelletizing plant, see *Skillings,* October 15, 1955, p. 2; November 26, 1955, p. 6; January 12, 1957, p. 6.

[16] See *Picture Magazine* of the *Minneapolis Sunday Tribune,* April 29, 1956; *Skillings,* April 7, 1956, p. 22. See also *Skillings,* January 12, 1957, p. 6, and March 2, 1957, p. 16, 17, for information on the 1956 shipments. 3,584,736 tons were shipped from Silver Bay before the close of the season on December 4, 1956. *Ed.*

[17] For pellet shipments by year, see Appendix 3, below.

[18] The members changed from time to time, but during this period, they usually were C. W. Beck and Ora E. Clark of Armco, and R. P. Liggett and John C. Murray of Republic.

[19] Among the many published descriptions of the dedication are those in *Skillings,* September 22, 1956, p. 7; *Duluth Herald,* September 12, 14, 1956; *Duluth News-Tribune,* September 14, 1956; *Minneapolis Star,* September 14, 17, 1956. *Ed.*

[20] The total cost of the completed project was about $187,000,000, roughly one per cent over Fred Darner's estimates — an unbelievably accurate estimate for a totally new project of this size.

[21] See Chapter 4, note 7.

[22] See *Wall Street Journal,* March 4, 1958.

[23] On the suit, see Reserve Mining Company *vs.* Mesabi Iron Company in 172 *Federal Supplement* 1–12 (District Court, Minnesota, 1959). A mimeographed copy of the *Trial Memorandum of Plaintiff, Reserve Mining Company* is in the collections of the Minnesota Historical Society. The Delaware cases are Putterman *vs.* Daveler *et al.,* 134/A 2d/480 (Delaware Chancery, 1957); on the motion to remove, see 169 *Federal Supplement* 125 (District Court, Delaware, 1958). See also *Silver Bay News,* June 24, 1958; *Minneapolis Sunday Tribune,* February 19, 1961. *Ed.*

[24] See *Wall Street Journal,* April 25, 1960.

[25] Copies of many speeches are on file in the public relations departments of Erie, Reserve, and Oliver. Chisholm spoke before the St. Paul Chamber of Commerce on December 15, 1953. See *St. Paul Pioneer Press* of that date. The Minnesota Historical Society has copies of talks given by Joseph S. Abdnor before the Lake Superior Chapter of the National Association of Accountants on September 20, 1961; by R. J. Linney before the Rotary Clubs of Duluth and Virginia on January 19, 1961, and February 7, 1962; by Donald Scott of Continental Sales and Equipment Company published in the Minnesota Section, American Institute of Mining and Metallurgical Engineers, *Twenty-third Annual Mining Symposium,* January 15–17, 1962, p. 121–131 (Minneapolis, 1962); and by Christian

F. Beukema before the Minnesota Press Association, as reported in *Mesabi Daily News*, September 26, 1960. For the material quoted below, see *Skillings*, October 20, 1951, p. 6.

CHAPTER 11 – THE REVOLUTION OF THE 1960S

[1] Haley, "Operating Results Using Taconite Pellets on Armco's Middletown Blast Furnace," in American Institute of Mining and Metallurgical Engineers, Blast Furnace, Coke Oven, and Raw Materials Committee, *Proceedings*, 20:15–31 (1961). Quoted material may be found on page 31. It is interesting to note that Kenneth R. Haley, the blast furnace operator who made the test for Armco, is the son of Kenneth M. Haley, who was then superintendent of Reserve's pelletizing plant and who is now that firm's manager of research and development. Following the announcement of this test, similar ones were made by other steel companies which confirmed and even improved upon the Armco results. See H. N. Lander and J. W. Banks, "Blast Furnace Operation with a One Hundred Percent Pellet Burden," in Minnesota Section, American Institute of Mining and Metallurgical Engineers, *Twenty-Fourth Annual Mining Symposium*, 1963, p. 37–45 (Minneapolis, 1963). The booklet contains a number of other papers of interest on "Taconite Developments."

[2] *Duluth Herald*, April 7, 1960; *Minneapolis Star*, August 4, 1960.

[3] A clean plant is a safe plant, and both Erie and Reserve are proud of their safety records. According to the company's records, from November, 1955, when Reserve's big project went into operation, to November, 1962, its employees have worked almost 29,500,000 man-hours with no fatalities and with only 77 accidents sufficiently severe to cause the injured employee to lose any working time. Statistically this is a frequency of only 2.62 lost-time injuries per million man-hours worked, compared to 9.66 for all concentration operations in the Lake Superior district. It is much safer to be employed in a taconite plant than it is to drive to work. L. J. Hall to the author, December 26, 1962. For published figures on Erie and partial ones on Reserve, see St. Louis County Inspector of Mines, *Annual Reports*, 1955 through 1963.

[4] On Cleveland-Cliffs here and below, see *Duluth News-Tribune*, January 9, 1955; J. S. Westwater, "Pelletizing by Grate-Kiln Method at Humboldt," in *Blast Furnace and Steel Plants*, June, 1961, p. 513–518, 530–535; *Skillings*, March 17, 1963, p. 6, March 14, 1964, p. 1, 4, May 16, 1964, p. 1.

[5] "World Iron Pellet Operations," a list compiled by the American Iron Ore Association, January 23, 1964; Otto G. Gramp, "Production of Iron Ore Pellets in North America," in *Skillings*, May 23, 1964, p. 4–6. On the proposed Oglebay Norton-Ford plant in Minnesota, see *Duluth Herald*, May 1, 1964; on United States Steel, see *Minneapolis Star*, March 5, 1963, *Duluth Herald*, September 12, 1963; on Hanna, see *American Metal Market*, April 9, 1964, and *Minneapolis Tribune*, June 17, 1964; on Jones and Laughlin, see *Duluth News-Tribune*, October 5, 1963.

[6] Walter A. Marting, "A Look at the Iron Ore Mining Industry," in *Skillings*, September 28, 1963, p. 1, 4; Gramp, in *Skillings*, May 23, 1964, p. 4. See also *Wall Street Journal*, January 23, 1963, January 28, 1964.

[7] On Republic, see *Minneapolis Tribune*, July 21, 1961; *Skillings*, September 28, 1963, p. 1. Figures for Pickands Mather were compiled from Alm, *Minnesota Mining Directory, 1962*. See also Donald W. Scott, "A Review and Appraisal of Iron Ore Beneficiation," in *Mining Congress Journal*, May, 1963, p. 56–59; June, 1963, p. 78–83; July, 1963, p. 53–56.

[8] American Iron Ore Association, "World Iron Pellet Operations," January 23, 1964.

[9] Figures for 1963 show 16,704,967 tons of taconite shipped for 10,636,395 man-hours worked, against 18,550,211 tons of natural ore shipped for 3,933,886 man-hours worked. See St. Louis County Inspector of Mines, *Annual Report, 1963,* p. 5, 9. Reserve's E. W. Davis Works is not in St. Louis County, and the 2,478,553 man-hours worked there in 1963 were added to the published figures in the inspector's report to arrive at these totals. The St. Louis County figures cover the Mesabi Range east of Keewatin and all of the Vermilion Range.

[10] Minnesota Natural Resources Council, *Natural Resources of Minnesota: 1962,* 49 (St. Paul, 1962). Total state and local taxes paid by Reserve in 1961 amounted to $2,486,000, according to Robert J. Linney, *Minnesota Taconite and Its New Competition,* 5, a booklet published by the company in 1962.

[11] Alm, *Minnesota Mining Directory, 1963,* 249, 253, 277.

[12] "How Soon Nuclear Blasting?" in *Engineering and Mining Journal,* March, 1964, p. 102; Alm, *Minnesota Mining Directory, 1963,* 249; Eugene P. Pfleider and Donald H. Yardley, "Underground Mining of Minnesota Taconite — A Future Probability," in *Twenty-Fourth Annual Mining Symposium,* 1963, p. 55–69.

[13] On semitaconite processing, see Appendix 2, below, and Allen D. Kennedy and John C. Nigro, "Roasting — Magnetic Separation — Flotation of a Michigan Fine Grained Iron Ore," in *Twenty-Fourth Mining Symposium,* 1963, p. 71–83. See also "Great Northern Ry.'s Taconite-Lignite Research Program," in *Skillings,* June 27, 1964, p. 1, 6, 21.

[14] It is possible in the laboratory to make pellets with a tumble test of 2 or 3 per cent, but there is some economic limit, and whether the cost of producing such superhard pellets can be justified remains to be seen. On the processing methods to make such superstrong pellets, see S. R. B. Cooke and R. E. Brandt, "Solid State Bonding in Iron Ore Pellets," in American Institute of Mining and Metallurgical Engineers, *Transactions,* 199:411–415 (New York, 1954).

[15] See Chapter 5, above.

[16] The proposed constitutional amendment is in *Session Laws,* 1963, p. 156. A discussion of the issues involved and the political background of the amendment may be found in a cogent booklet entitled *Proposed Amendments to the Minnesota Constitution 1964,* issued by the League of Women Voters of Minnesota (15 pages). Reactions to the 1961 proposal are reflected in a booklet entitled *The Taconite Issue and Minnesota's Press* (n.p., n.d.), a copy of which is in the collections of the Minnesota Historical Society.

[17] The author was appointed honorary chairman of this committee; its chairman is Dr. Charles W. Mayo. *Ed.*

Index

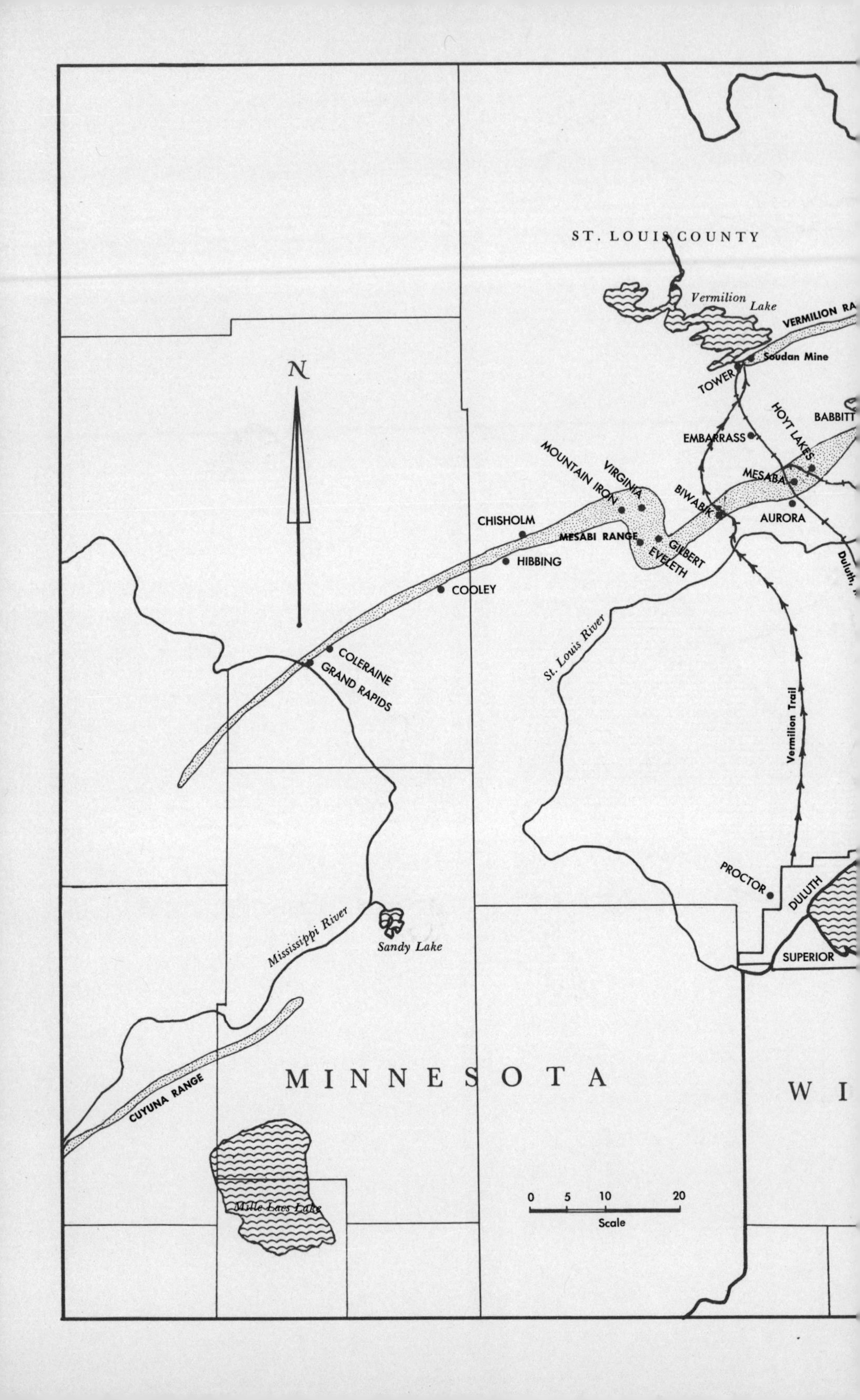

ST. LOUIS COUNTY
Vermilion Lake
VERMILION RA
Soudan Mine
TOWER
HOYT LAKES
BABBITT
EMBARRASS
MOUNTAIN IRON
VIRGINIA
MESABA
BIWABIK
AURORA
CHISHOLM
MESABI RANGE
GILBERT
EVELETH
HIBBING
COOLEY
N
St. Louis River
COLERAINE
GRAND RAPIDS
Vermilion Trail
PROCTOR
DULUTH
SUPERIOR
Mississippi River
Sandy Lake
MINNESOTA
W I
CUYUNA RANGE
0 5 10 20
Scale
Mille Lacs Lake